THE POPE OF PHYSICS

THE POPE
OF
PHYSICS

ENRICO FERMI AND THE
BIRTH OF THE ATOMIC AGE

GINO SEGRÈ AND **BETTINA HOERLIN**

HENRY HOLT AND COMPANY   NEW YORK

Henry Holt and Company
Publishers since 1866
175 Fifth Avenue
New York, New York 10010
www.henryholt.com

Henry Holt® and 🅷® are registered trademarks of
Macmillan Publishing Group, LLC.

Library of Congress Cataloging-in-Publication Data

Names: Segrè, Gino, author. | Hoerlin, Bettina, 1939–author.
Title: The Pope of Physics : Enrico Fermi and the birth of the atomic age /
    Gino Segrè and Bettina Hoerlin.
Description: First edition. | New York : Henry Holt and Company, [2016] |
    Includes bibliographical references and index.
Identifiers: LCCN 2016013398| ISBN 9781627790055 | ISBN 1627790055 |
    ISBN 9781627790062 | ISBN 1627790063
Subjects: LCSH: Fermi, Enrico, 1901–1954. | Physicists—Italy—Biography. |
    Physicists—United States—Biography.
Classification: LCC QC16.F46 S44 2016 | DDC 530.092 [B]—dc23
LC record available at https://lccn.loc.gov/2016013398

Our books may be purchased in bulk for promotional, educational, or business use. Please contact
your local bookseller or the Macmillan Corporate and Premium Sales Department at (800) 221-7945,
extension 5442, or by e-mail at MacmillanSpecialMarkets@macmillan.com.

First Edition 2016

Designed by Kelly S. Too

Printed in the United States of America

1   3   5   7   9   10   8   6   4   2

To immigrants, then and now

CONTENTS

Prologue: Trinity   1

PART 1: ITALY, BEGINNINGS

1. Family Roots   7
2. The Little Match (Il Piccolo Fiammifero)   13
3. Leaning In: Physics and Pisa   18
4. Student Days   24
5. The Young Protégé   28
6. The Summer of 1924   35
7. Florence   41
8. Quantum Leaps   49
9. Enrico and Laura   56

PART 2: PASSAGES

10. The Boys of Via Panisperna   65
11. The Royal Academy   70
12. Crossing the Atlantic   76
13. Bombarding the Nucleus   81

14. Decay 87
15. The Neutron Comes to Rome 93
16. The Rise and Fall of the Boys 102
17. Transitions 110
18. Stockholm Calls 116

PART 3: HELLO, AMERICA

19. Fission 127
20. News Travels 135
21. Chain Reaction 144
22. The Race Begins 151
23. New Americans 159
24. The Sleeping Giant 167
25. Chicago Bound 176
26. Critical Pile (CP-1) 184
27. The Day the Atomic Age Was Born 193

PART 4: THE ATOMIC CITY

28. The Manhattan Project: A Three-Legged Stool 201
29. Signor Fermi Becomes Mister Farmer 209
30. Götterdämmerung 219
31. The Hill 226
32. "No Acceptable Alternative" 234
33. Aftershock 244
34. Goodbye, Mr. Farmer 253

PART 5: HOME

35. Physicist with a Capital *F* 261
36. The Fermi Method 268
37. The Super 275
38. Circling Back 283
39. Last Gift to Italy (*Ultimo Regalo all'Italia*) 291
40. Farewell to the Navigator 296

Afterword   303
Notes   309
Bibliography   326
Acknowledgments   334
Index   337

THE POPE OF PHYSICS

TRINITY

July 16, 1945. Dawn broke reluctantly, the early rays of the day barely grazing the tops of nearby peaks. It was almost as if the sun sensed that its brightness would be outshone. A group of scientists, huddled together against the morning chill, set aside their worries about bad weather and concentrated on the seismic event that was about to occur.

The countdown began at 5:09:45 a.m. If all went according to plan, exactly twenty minutes later they would throw a switch triggering the detonation of the world's first atom bomb. The tension at the site was palpable as they waited. An earthen mound above a concrete slab roof, all supported by massive oak beams, fortified the structure they were staying in. Located ten thousand yards south of the hundred-foot-high tower holding the bomb, the shelter was thought to be safe no matter how large the explosion at Ground Zero might be. The small group who manned the Trinity Project, as it was named, included George Kistiakowsky, the head of explosives, Kenneth Bainbridge, the man who had selected and built up the site, and of course J. Robert Oppenheimer.

General Leslie Groves, feeling that he and Oppenheimer should not be together in case of disaster, had gotten into his jeep a little earlier and driven five miles south to Base Camp, leaving his military deputy in charge at the bunker. Most of the physicists who had worked on Trinity were at Campania Hill, some twenty miles northwest of Ground Zero. A few, including Enrico Fermi and Emilio Segrè, were ten miles closer, at Base Camp. Shallow trenches had been dug there to protect them, but would those trenches be enough? Everybody thought yes, but how big was the blast going to be? Might it even be a complete failure?

A few days earlier the senior physicists had started a betting pool about the blast's magnitude; a one-dollar entrance fee built up the pot. Kistiakowsky's wager had been one thousand tons of TNT equivalent, a low estimate, as he would discover when he climbed on top of the bunker after the blast, only to be knocked over by the shock wave that reached him a few seconds later. Hans Bethe, head of the theory division, had said eight thousand, while a worried Oppenheimer had settled for a modest three hundred.

The switch was thrown at 5:29:45. Many recorded their impressions of what happened next, an event subsequently described as brighter than a thousand suns. From base camp, Isidor Rabi's memory was that "Suddenly there was an enormous flash of light, the brightest light I have ever seen or that I think anyone has ever seen. It blasted, it pounced: it bored its way right through you. It was a vision that was seen with more than the eye." The flash was so overwhelmingly intense that irrational instants of fear were common. "For a moment I thought the explosion might set fire to the atmosphere and thus finish the earth, even though I knew this was not possible," Segrè recalled.

Seconds later, as the mushroom cloud began rising in the sky, those watching it were left trying to grasp the meaning of what they were witnessing. Oppenheimer remembered the lines from the *Bhagavad Gita*'s scriptures coming to him: "I am become Death, the destroyer of worlds." Bainbridge expressed himself in much more prosaic language: "Now we are all sons-of-bitches."

Fermi was arguably the physicist most responsible for the world-changing event that had just occurred in the New Mexico desert. There is no documentation of what he was thinking at the time. But there is a record of what Fermi was doing. If you didn't know him, it would seem bizarre, but everybody knew he always acted with purpose. A few seconds after the blast, Fermi stood up and began tearing a large sheet of paper into small pieces and then dropping them from his upraised hand. Forty seconds later, as the front of the shock wave hit, the mid-air pieces were blown a short distance away. Pacing off the distance to where they landed, some eight feet, he consulted a little chart he had prepared beforehand. Shortly afterward, Fermi told those around him that he estimated the blast's force as roughly equivalent to ten kilotons of TNT.

A few hours later, Fermi climbed into a special lead-lined tank and headed toward Ground Zero to scoop up material for a more careful assessment of what had transpired. The detailed measurements took about a week. It was concluded that the blast's magnitude, corresponding to twenty kilotons of TNT, was close to the estimate he had made within a minute of the explosion. None of the physicists was surprised.

The dropping of the paper pieces soon became yet another vintage Fermi story, adding to the lore of how he could, with the simplest of means, estimate the magnitude of any physical phenomenon. And, as usual, he had been right. His colleagues in Rome used to joke that Fermi was infallible, like the Pope. He had acquired the nickname "the Pope of Physics" early on. It was an appellation that persisted, deservedly, throughout Fermi's lifetime.

PART 1

ITALY, BEGINNINGS

1

FAMILY ROOTS

Enrico Fermi's ancestral roots can be traced to the valley of Italy's greatest river, the Po, whose origins lie in the western Alps. It flows from west to east, neatly bisecting Northern Italy, and finally empties into the Adriatic Sea. As it travels its four-hundred-mile course, it grows steadily in volume, fed by rivers coursing down from the Alps and from ones born in the central Apennine mountain chain.

The Po Valley, defined by its river, is agriculturally fertile and culturally vibrant. It is also Italy's economic center, thanks to large industries, but enriched by a wealth of small businesses that have adapted an older tradition of craftsmanship to the demands of the new commerce. Turin, the automotive home of Fiat, is located directly on the river. As it winds further, Milan, a center of style and fashion, is a little north, and Bologna, known for its culinary treats, a little south, of its course. Venice, an architectural wonder, is not far from the delta where the Po empties into the sea. These are the region's dominant cities, but there are a number of midsized ones with their own history and institutions.

In most cases this diversity comes from their ancient founding during the Roman Empire and sometimes even earlier, followed by an evolution during the Renaissance into independent city-states. What we currently call Italy, a country that came together only in 1870, was until shortly before then little more than a hodgepodge of smaller fiefdoms wavering opportunistically to and fro in their allegiances to larger European powers.

Piacenza, the ancestral home of the Fermis, lies in the midst of the Po Valley. It has a particularly impressive thirteenth-century city hall, but the town is largely neglected as a tourist attraction because it lies almost directly in the middle of a triangle formed by the better-known Parma, Cremona, and Pavia. Founded by the Romans in 218 B.C.E., the settlement was given the name Placentia, from the Latin *placere*, "to please." Through the following centuries, it did indeed please, and in doing so underwent the same cyclical sacking and rebuilding that its neighbors suffered.

In 1545 the Duchy of Parma and Piacenza was established. Except for a brief interlude during Napoleon's short-lived conquest of Northern Italy, it controlled the region surrounding those two cities until the formation of modern Italy. Shortly before that, the Fermis, a local family, made the transition away from tilling the soil. The man who would become Enrico's grandfather, Stefano Fermi, entered the employment of the state and rose to be administrative head of a small municipality adjacent to Piacenza.

Enrico Fermi's grandfather married Giulia Bergonzi, a woman thirteen years his junior, and began a large family; their second son, Alberto, would one day become Enrico's father. The fluid national identities that characterized the Italian peninsula in the first years after Alberto's birth in 1857 were such that he entered the world as a subject of the Duke of Parma and Piacenza, became a resident of the free territory of Emilia two years later and a citizen of the Kingdom of Sardinia a year after that, and finally, at age four, became an Italian. All this without ever leaving the vicinity of Piacenza.

In the 1840s, Alberto's father, Stefano, had settled with his wife in Caorso, a small municipality that lay eight miles east of Piacenza. Their life was a simple one, centering on family, work, and church. He and his wife, Giulia, almost certainly went occasionally to Piacenza, but they probably did not venture as far as Cremona, for though it was only eight miles west of Caorso, going there required crossing the Po and entering a different country.

Those borders vanished in 1861 with the emergence of the Kingdom of Italy. Stefano and Giulia hoped there would be opportunities for advancement in this new country, still woefully underdeveloped by comparison to those of Northern Europe. The Industrial Revolution had for all practical purposes bypassed the peninsula, and most of Italy's workers either labored on the land as they had for centuries or engaged in minor commerce. Nor was transportation very different from what it had been since Roman times, for there was little more than fifteen hundred miles of rail lines in the whole country, almost all of it north of the Po.

Education was viewed as providing a first step to bettering oneself. More than three quarters of Italy's population was still functionally illiterate. Many could read a little but, like Enrico's grandmother Giulia, had not learned to write, much less how to deal with arithmetic problems beyond a simple shopping list.

The newly formed Italian government instituted a set of reforms designed to change this state of affairs. An innovative law called for universal enrollment in elementary school starting at age six. Attendance in the first four years was compulsory, though in practice the rule was often broken. The poor considered it a luxury to have their offspring removed from the workforce. The rich educated theirs at home.

Stefano and Giulia, despite their modest circumstances, insisted on having their children attend school, and Alberto, who seemed to be the most scholastically gifted of them, advanced beyond an elementary education. But given the Fermi family's financial situation, attending university was never considered. When he reached the age of sixteen, Alberto's schooling was finished and it was time for him to seek employment.

By then Rome was the capital of Italy. The city and its surrounding region had been an independent state under papal rule until 1870, when the Kingdom of Italy had annexed the territory. Pope Pius IX had declared the occupation violent, unjust, and invalid. Retreating into the Vatican, he refused to recognize the existence, much less the legitimacy, of the new Italy.

Alberto Fermi, thirteen at the time, must have followed the story with keen interest. His parents, especially his mother, were devout Catholics, but he was already having the doubts that would later turn him into an agnostic, if not an atheist.

Alberto knew that he would have to leave Caorso if he was to advance in the world. Working for a company that built and managed rail lines seemed to be a particularly interesting choice in the early 1870s. At the time of his birth Italy had more than two dozen independent railway companies, each operating separate lines. Many of the lines had been founded with foreign capital, making them dependent on events outside Italy's control. Each attempted to maximize its profits with no concern for helping to forge a national identity.

By the age of twenty-four, Alberto was employed in the service of the company that managed the northern Italian railroads, one of the four that had emerged from consolidation. Through various reorganizations, the railroads continued to employ him until his retirement. In 1905 he became a civil servant for the Italian railroads, nationalized and combined into a single company, the Ferrovie dello Stato.

During all the years of employment, Alberto's willingness to work harder than anybody else, combined with his organizational ability, perseverance, and native intelligence, had led to his steady rise in the ranks. These personality traits would be very much imprinted on his only surviving son, Enrico.

Like his father, Alberto did not marry until he was forty-one. Ida de Gattis was fourteen years his junior. The daughter of an army officer, she was born in Bari, a city in Puglia, commonly known as the heel of Italy. Ida had been orphaned at a young age and raised by relatives in Milan. Like Alberto, she strongly advocated self-sufficiency and self-reliance. She began teaching after a three-year course for elementary school teachers, her ambitious trajectory a relative rarity at a time when women were still discouraged from entering a profession.

Ida and Alberto were both intelligent and upwardly mobile. They were not cultured in the traditional sense of appreciating art, music, and literature, though Alberto, a rather taciturn man, was known to occasionally break into song in the privacy of his home when shaving or bathing. His choice was almost always a Verdi aria, probably because the composer was born in Busseto, a small town only a few miles from Piacenza.

When Ida and Alberto married, they settled in Rome on Via Gaeta, a street that lay a short distance from the central railroad station. Their apartment was in one of the newer buildings that had sprung up during the thirty years since Italian unification, a period during which the city's population had roughly doubled to some four hundred thousand. Their neighbors were much like the Fermis, upwardly mobile middle-class people, the husbands typically employees of the government or of a quasigovernmental agency.

The Fermis lived on Via Gaeta for ten years, then moved to a nearby apartment in 1908. Slightly more spacious, but still far from luxurious, it had no central heating and its bathroom, as was not unusual at that time, was equipped with only a sink and a toilet. Baths were taken in two zinc tubs, a smaller one for children and a larger one on casters for the parents. By then Ida and Alberto had three children. Maria was born in 1899, Giulio in 1900, and Enrico on the twenty-ninth of September, 1901.

The closeness in age of the children and Ida's desire to continue teaching resulted in Enrico's being placed in a farm family. The tradition of having wet nurses for infants was centuries old in Italy, usually only adopted by the upper classes. A young woman who had recently given birth would be brought in from the countryside to nurse and care for the baby, living with the family at least until the child was weaned.

Toward the end of the nineteenth century, the reverse was becoming common for middle-class couples living in big cities: their children were sent to the countryside. With three children in less than three years, the Fermis made such arrangements for Enrico, the youngest. At the time farms still existed close to Rome; it was not too hard to find a suitable family willing—for a fee, of course—to take on a little boy for a few years.

A child psychologist, ruminating on where such a beginning would lead, might conclude that in adulthood the person would be either very self-reliant and controlled or overly needy and dependent. Enrico was obviously an example of the former.

One can speculate that the farm family provided a loving

environment and a place where he could observe, explore, and enjoy nature. The security that Fermi exuded, as well as his love of the outdoors, might be related to those farm years. Yet the pain of separation from his family of birth must have affected his development, too, and is probably related to why Fermi kept emotions to himself and never complained. This is how he learned to cope.

And those coping skills served him well in later life.

2

THE LITTLE MATCH

(Il Piccolo Fiammifero)

Enrico's sister remembers him being "small, dark and frail-looking" when he rejoined the family at age two and a half. She also recalled how, probably worried by the sudden appearance of all the strangers, he immediately began crying, only to be told by his mother "to stop at once; in this home naughty boys are not tolerated." Little Enrico did stop crying. But with bottled-up frustrations, he was known to occasionally break into flaming rages, earning him the family nickname of Piccolo Fiammifero (Little Match).

Rages were not tolerated any more than an untidy appearance. Perhaps rules had been slacker in a farm environment. In a city setting, his mother apparently would insist that his face always be clean, stopping at fountains to wash up during their excursions. Despite Alberto and Ida's strictness in child rearing, the Fermi family appears to have been close, with a special bond developing between the two brothers, Giulio and Enrico, only a year apart in age.

Alberto and Ida had advanced in their careers despite not having university degrees; but, like many upwardly mobile parents, they wanted more for their children. All three of those children, through a combination of natural ability and the self-discipline learned from their parents, excelled in their studies, each of them consistently at the top of his or her class.

At the beginning of the twentieth century the advanced school curriculum was still oriented toward a classical education, emphasizing Latin and Italian literature with Greek added in the final five years. Mathematics, history, and science were also taught, but regarded as secondary. In particular a student preparing for the culminating graduation exam, the so-called Maturità, was expected to know practically by heart Dante's *Divina Commedia*, Italy's national literary treasure. Fermi had little interest in music or visual arts, but a love of poetry, and not only Dante, remained with him. On long hikes he could occasionally be heard reciting under his breath verses he had learned in his youth.

Maria, who eventually would become a high-school literature teacher, was drawn to the humanistic side of the studies. By contrast, Giulio and Enrico were more interested in science or at least in the technical skills they acquired by building models and little electrical motors.

The first recognition that Enrico was exceptional occurred shortly after he turned thirteen. Alberto Fermi, nearing sixty, had continued to rise in his profession and had become a chief inspector in the Ministry of Maritime and Railroads. Its offices were located in a building a little less than a mile from the Fermi apartment, and Enrico had started meeting his father at the building's door after work and walking home with him. Adolfo Amidei, a thirty-seven-year-old engineer employed in the same office as Enrico's father, often joined them for part of the way since his apartment was in the same direction as theirs.

When Enrico discovered that Amidei had an interest in mathematics, he asked him a few questions about geometry that his father had been unable to answer. To help his colleague's son, Amidei lent Enrico a geometry book. The young boy quickly worked out the solutions to the problems it contained, some of which even Amidei had not been able to solve. Impressed, Amidei inquired of his senior colleague whether anybody else had commented on his son's skill and precocity. Alberto told him that Enrico had always done well in school, but none of his teachers had noted anything out of the ordinary.

At about the same time, early January 1915, tragedy struck the family. Giulio had developed a throat abscess that was interfering with his breathing, not an uncommon consequence of a severe tonsil infection. Today, treatment is by massive doses of antibiotics, thereby usually avoiding any surgical intervention. The standard procedure in 1915 was an incision and drainage of the abscess under a local anesthetic. This surgery was performed on Giulio in a clinic, his mother and sister waiting to take him home after the anesthetic wore off. But Giulio had a dramatic adverse reaction to its administration, went into anaphylactic shock, and died on the operating table.

The family was devastated. Alberto became even more taciturn and Ida fell into a deep depression; Giulio, warmer and more outgoing than Enrico, had been her favorite. Ida's fits of inconsolable crying lasted for hours and she was in no shape to help alleviate the pain of others. Enrico was left to grieve on his own. To prove to himself that he was not a total wreck, a week later he deliberately walked by the clinic where his brother died. It was a striking example of Fermi's early need for emotional control.

One way for thirteen-year-old Enrico to fill the traumatic and heartbreaking void was by hard work. Amidei, recognizing both the boy's loneliness and his eagerness to learn, tried to do what he could by continuing to lend him books; the more he did, the more impressed he was by Enrico's intelligence and thoroughness. When Amidei once asked his young protégé if he wanted to keep a calculus book he had lent him, he was told that it wasn't necessary because he had thoroughly assimilated the material. As people would repeatedly say over the next forty years, "When Fermi knew something, he really knew it."

Another precious balm for Enrico was having a schoolmate of Giulio's become his close friend. Enrico Persico shared his interests in science and in building mechanical objects. The two of them soon began taking long walks together, discussing their common dreams. Fifteen years later, the two Enricos would be Italy's first two professors of theoretical physics. And almost forty years later, they would still take long walks together and share dreams.

The young Fermi, insatiable in his quest to learn more about science, found his first real physics book in Rome's Campo dei Fiori (literally Field of Flowers). This square, located between the Tiber and the ruins of Pompey's Theatre, the site of Caesar's assassination, remains to this day one of Rome's liveliest areas. Now a popular outdoor food market, during Enrico's childhood it was a horse market two days a week and once a week held stalls where one could purchase new and old books. Most of them were novels or, this being Rome, theological treatises. Occasionally one might find something else.

One day in late 1915, the two Enricos were looking through the collections in the Campo. Fermi picked up a nine-hundred-page two-volume set entitled *Elementorum Physicae Mathematica*. It was a text on mathematical physics written in the 1830s by a priest named Andrea Caraffa who had taught science and mathematics at the Collegio Romano, a Roman university founded in the sixteenth century by the Jesuit order. With more than four years of intense study of the language completed, Enrico had no problem with the Latin text. In any case, all the equations were in the universal language of mathematics.

Purchasing the book with his allowance, Enrico studied it carefully in the weeks that followed, making notes when he had questions. Caraffa's subject matter was early-nineteenth-century physics, chiefly the mechanics of celestial motion and wave theory. The mathematics Caraffa employed might have provided an insurmountable obstacle had Enrico not already studied the subject on his own with books lent by Amidei. Not only was Fermi the only twentieth-century physics genius to be entirely self-taught, he surely must be the only one whose first acquaintance with the subject was through a book in Latin.

At sixteen, Enrico skipped his final year of high school. It was time for him to begin thinking about what would come next. Obviously headed toward a university, Enrico was expected to stay at home, since Italian universities had no dormitories. But Amidei felt his young protégé would benefit greatly from being away from the oppressive atmosphere that dominated the Fermi household after Giulio's death.

Once again Amidei's important role in Fermi's development surfaced. He was familiar with an elite institution in Pisa, the Scuola Normale Superiore. Admission to its small entering class, some forty students in all, was by competition. Amidei was confident that Enrico would shine. Since the school provided room and board for its matric-

ulating students, there would be no additional financial burden on the
Fermi family. The classes were mostly big lectures held at the University of Pisa, but additional supervision and instruction would be available in the Scuola. Such an opportunity would enrich Enrico both intellectually and emotionally.

Discreetly inquiring of Enrico whether he would like to attend, Amidei received an enthusiastic response. He next set out to speak to the boy's parents. They were reluctant, particularly Ida, who felt she would be losing her other son. But Amidei was persuasive in convincing her and Alberto of the benefits to Enrico. He stressed that attending Italy's premier institution of higher learning, a school from which many of the country's most famous scholars, political figures, and writers had graduated, would open many doors for him. Ida and Alberto finally agreed to let him apply for admission.

Amidei had also encouraged Enrico to study German. The boy learned French, part of the usual school curriculum, but an increasing body of the scientific literature was in German, and knowledge of that language would stand Enrico in good stead. It was an interesting although unpopular suggestion, since Italy was at war with Austria and Germany.

When World War I began in August 1914, Italy initially opted to remain neutral. This probably would have been the wisest policy to continue and was generally favored by the population at large. However, in the spring of 1915, the Italian prime minister began secret negotiations with the French and British to enter the war on their side. He stoked popular opinion to this end, and in May 1915, Italy declared war on Austria. The Italian army suffered a series of defeats, climaxing in October 1917. During the closing weeks of the war, the Italian army managed to garner a significant victory against a demoralized Austrian army.

Fortunately, Enrico Fermi had still been too young to be drafted into the army. Less than two weeks after the armistice was signed, he took the entrance examination for the Scuola Normale.

3

LEANING IN:
PHYSICS AND PISA

Though a superb mathematics student, Enrico Fermi selected physics as his field of study. One of the reasons was his affinity, since childhood, for performing experiments and his fascination with the apparatus they required. His interest in science had started with models and little motors built with his brother. After Giulio's death he continued in the same vein with Persico, the other Enrico. They gradually embarked on more sophisticated ventures: accurate measurements of gravity's acceleration, water's density, and the level of atmospheric pressure.

Part of their work was carried out at the Rome Meteorological Institute, a center Fermi had become familiar with because its director had been his high school science teacher, a man who helped him and Persico build barometers. At the center's library Fermi found another influential book, denser and more up-to-date than the one he had purchased at Campo dei Fiori. This time French, not Latin, was the tome's language.

A four-volume five-thousand-page encyclopedic treatise on physics, it was then popular throughout Europe. It was written by a Russian physicist, Orest Khvolson, and had been translated into several languages, though not into Italian. Covering physical phenomena as well as state-of-the-art instruments, it is remarkably thorough even by today's standards. Fermi studied Khvolson at a rapid clip in the summer of 1918, laying the foundation for his astonishing command of every aspect of classical physics. In letters to Persico, he described going through the book at more than a hundred pages a day, skipping large sections of material he was already familiar with.

At the end of August, when Fermi was seventeen years old, the announcement came for admission exams to the Scuola Normale, offered on four successive days starting October 28. Several hundred candidates were expected to apply for the few dozen openings available in all fields. No more than a handful of prospective future mathematicians and physicists would gain admission. Fermi's acceptance by the Scuola, despite Amidei's faith in him, was not a foregone conclusion.

The exams were extensive. The first three days consisted of eight-hour written exams, and the fourth would be an oral one. Candidates for admission in physics and mathematics would be tested on their knowledge of algebra, geometry, and physics, each day's exam positing one problem and one essay on a theme unknown to the candidates beforehand.

The entrance examinations to the Scuola were postponed indefinitely on account of an influenza epidemic. The atmosphere in Rome that autumn, already grim because of the war, turned darker. However, by early November, the number of flu cases diminished, the war seemed to be drawing to a close, and the upcoming academic year was starting. The decision to delay was reversed and the examination rescheduled for November 12. Though the schedulers could not have known it at the time, the armistice would be signed one day earlier.

Fermi performed brilliantly in the first two days of examinations, but his work on the third day is what led to his designation as a precocious genius. The topic for the physics essay was "Distinctive Characteristics of Sound and Their Causes." A natural starting point was to consider how a vibrating string leads to sound waves propagating in air. Fermi displayed his bravura by going well beyond this scenario. He tackled the much harder problem of how the vibrations of a rod attached to a wall generate sound waves.

Fermi derived the equation describing the motion of the rod and proceeded in masterly fashion to solve it. It was an essay that one could barely imagine a talented graduate student being able to write. It was unthinkable to see it from the pen of a self-taught high school student.

The head of the three-man committee administering the examination in Rome was Giulio Pittarelli, a distinguished geometry professor at the University of Rome. Pittarelli, an unlikely person to break academic traditions, was well aware that communication with competing students was disapproved until exams were over and the Scuola Normale had decided whom to accept. Nonetheless, he could not restrain himself. He had been so impressed by Enrico's essay that he called him in to his office.

Quaking at the request of one of his examiners to see him privately, Fermi entered the dimly lit room. The boy, having just turned seventeen, stood before the sixty-six-year-old professor and was subjected to a round of questions. After assuring himself that Fermi had understood everything he had written, Pittarelli informed him that he definitely would be admitted. It was unimaginable that any other candidate could do as well. Pittarelli also added that in forty years of teaching he had never encountered so gifted a student.

This was a tremendous boost to Fermi's self-confidence. Neither his parents nor his schoolteachers had thought there was anything truly exceptional about him. Amidei had, but he was a family friend and a modest man, not someone who was in a position to compare him to other would-be scientists. But Pittarelli had long been exposed to outstanding students, and was a university professor. He was going far out of his way to praise him, a fact for which Fermi would always be grateful.

The examining committee awarded Fermi the highest possible grades in all his examinations. They unanimously recommended admission to the Scuola. Fermi was thrilled to begin a new phase of life— and to do so in the birthplace of Galileo. This city was the cradle of physics, where the great master was born, studied, and began his teaching career. Galileo had deduced the laws of motion by measuring the periods of a swinging pendulum in the Duomo and by observing objects falling from the Leaning Tower. The subject of physics had been created in his image: experiment, observe, and deduce.

When Fermi arrived in Pisa in early December of 1918, it was a rather sleepy town of 65,000 inhabitants with signs of a glorious history but little else. The university had persevered over the centuries, as had the Scuola Normale affiliated with it. Fermi had taken the four-hour train ride from Rome to Pisa, descending at the station on a cold but sunny day with his two suitcases—one with clothes, the other with a few books and things for his room. A ten-minute walk brought him to the Arno River, smaller than he had imagined it would be. Crossing it on the Ponte di Mezzo, he entered the medieval part of the city with remnants of its circuit of old walls. A short walk along the Borgo Stretto soon led him to the glorious Piazza dei Cavalieri, the square that held the sixteenth-century Palazzo della Carovana where the school was housed.

A porter showed Enrico to his room. It had a marvelous view of the square and, in the distance, the Leaning Tower. Like the other rooms for the forty or so students, it was small and monastic. It held only a bed, a table, a chair, a shelf, and a sink. There was no running hot water and no heating other than the hand-held ceramic brazier that could be filled with hot coals. Enrico did not regard this as hardship since it was what he was used to in Rome. As a boy he had studied in the evening sitting on his hands to keep them warm, turning pages with the tip of his tongue.

Persico, who had taken the more conventional route of enrolling at the university in Rome, received a postcard from his friend within a few days. In it Fermi confessed to early pangs of homesickness, adding that he had quickly overcome them: "During the first days of the new life, I was slightly despondent. However everything has now passed and I have completely regained my self-control." He could always confide in Persico in a more intimate way. Persico was almost the brother he had lost, and their closeness, intense during their teens and early twenties, would persist throughout Fermi's life.

Fermi was one of a handful of physics students in Pisa. At the time, the subject was not highly regarded in Italy, nor did anyone seem to notice the great advances taking place in the field elsewhere. As he would find during the next two years, Fermi was the only person in Pisa to have any real understanding—much less an appreciation—of Max Planck's 1900 introduction of the quantum, Albert Einstein's general theory of relativity, and Niels Bohr's 1913 model of the atom. Physics in Italy was regarded as a purely experimental subject and taught accordingly, with emphasis on phenomena adaptable to simple classroom demonstrations.

By contrast, mathematics was firmly established in the country. Italy's best researchers in the field, very much up to date, had frequent and fruitful exchanges with colleagues from abroad and taught their students the latest developments. Mathematics' primacy was reflected in the apportioning of university professorships. Pisa had five mathematics professorships while physics had only a single one. Rome was the only Italian university in which there was more than one physics professor—there were two. University courses in mathematical physics were offered, but these basically focused on using physical phenomena such as planetary motion to analyze mathematical structures.

The lone Pisa physics professor, Luigi Puccianti, in his midforties, was a kindly, friendly man. He had made some notable investigations in his youth, sufficient to warrant his appointment in Pisa. Since then, he had given up on research, limiting his activities to teaching aided by an assistant who had graduated from the Scuola only a year earlier. There was little chance that either of them would be able to teach Fermi anything, but at least they did not perceive his clear superiority as a threat to their own position. To the contrary, in the following years, Puccianti and his assistant would often ask him to explain questions in modern physics they had difficulty understanding. They were also appreciative of the fact that Fermi had chosen physics, not mathematics, as a career path. Having someone so talented opt for their discipline would confirm the subject's importance.

A question naturally arises: How did Fermi learn all the consequential physics of the day? He did it the same way he had prepared for admission to the Scuola Normale: by studying relevant books. In his first two years in Pisa he read and completely absorbed contemporary texts in French, German, and English. The list includes Poincaré's

Théorie des Tourbillons, Sommerfeld's *Atombau und Spektralinien*, Rutherford's *Radioactive Substances and Their Radiation*, and several others. Beyond that, he began consulting recent major journals, in particular the German *Zeitschrift für Physik* subscribed to by the university's library. With his background in German, Fermi was probably the only one in Pisa to read them.

Once Fermi had absorbed the contents of what he had read, he neatly transcribed what he deemed the essential part of a complicated argument, often just a few annotated equations, into little notebooks. This lifelong habit allowed him to retrieve key notions in a way that appeared miraculous. Among the most moving are his teenage notebooks, filled with formulas and exercises, composed even before he entered the Scuola Normale and now stored in Pisa's Domus Galileiana collection.

While Fermi studied hard at the Scuola, he fortunately found a friend with whom he could have fun. Other than his serious-minded albeit close relationship with Persico, the element of youthful mischief and camaraderie had been missing from his life. Happily, he came to experience that in Pisa.

4

STUDENT DAYS

At the beginning of the second month of required university courses at Pisa, Fermi found himself sitting next to a tall, thin first-year engineering student. The two of them began chatting about what university life was like, what they expected to learn, and what they already knew. Many years later that student, whose name was Franco Rasetti, would remember what he told his mother after the encounter, "I have met another student who is a genius, a man like I have never seen before. He must be a sort of prodigy. He knows more than all the professors put together in physics and he understands everything." Rasetti would become Fermi's close collaborator over the next twenty years.

This only child was in many ways as exceptional as Fermi. Educated at home through elementary school by a father who was an avid naturalist and a mother who was a gifted painter, Rasetti had benefited greatly from both their talents. At a young age he had already collected enormous numbers of insects, animals, and plants, all of whose Latin names he knew, and he had provided extraordinarily accurate pictorial representations of the samples, even publishing articles in the *Bulletin of the Italian Entomological Society*. And that was not all: with an omnivorous appetite for knowledge and a phenomenal memory, he had read widely in several languages and taught himself chemistry.

On entering university, Rasetti chose engineering as a field, having decided that this would allow him to have a comfortable and stable career. He set aside his other passions, avocations to be indulged in periodically. Meeting Fermi changed Rasetti's life. Soon the young Pisan became convinced by his new friend that physics was a field of tremendous interest, where his extraordinary experimental dexterity would be valued. The two became inseparable, Fermi often eating dinner at Rasetti's home, where he lived with his parents. The family meal provided welcome relief from the dull fare Fermi encountered at the Scuola.

In turn, Rasetti had a great influence on Fermi. In addition to providing collegial company, Rasetti introduced his urban peer to the beauties of the Apuanian Alps, a nearby chain of six-thousand-foot peaks. Located along the coast some twenty miles north of Pisa, these scenic mountains are famous for the views of the sea and for the many quarries located near the central town of Carrara. These quarries are the source for marble used in the construction of some of the world's most striking buildings, particularly Italy's splendid churches.

The mountain peaks also allow for technical climbing. Rasetti had been initiated into the sport as a teenager, had loved it, and had made some noteworthy Alpine ascents. Fermi did not share Rasetti's desire to summit steep mountains on treacherous and challenging routes, but hiking was much to his liking. He did not have any trouble keeping up with Rasetti. Like all the male Fermis, Enrico was somewhat on the short side, with broad shoulders and a torso that was a little long in comparison to his legs. But those sturdy legs never seemed to tire, and Fermi would not hesitate to start out on a fifteen-mile excursion with a heavy pack on his back.

For Fermi, Rasetti also provided a release from his gloomy home atmosphere in Rome. Belatedly, Enrico was able to indulge in the adventures and harmless pranks of youth. He and Rasetti formed a two-person club for playing practical jokes, calling themselves the Anti-Prossimo, or Anti-Neighbor Society. One of their simplest tricks was to walk by one of the outdoor urinals common in Italy at that time and inconspicuously throw a small piece of sodium between the feet of the unsuspecting man using the facility. The resulting geyser of foam would predictably alarm the user, much to the amusement of the pranksters.

Occasionally Fermi and Rasetti brought their mischievous ways to the classroom. Once, when a lecturer was scheduled to explain how cats manage to turn themselves around during a fall to land on their feet, the Anti-Prossimo decided to sneak a live cat into the lecture hall and throw it into the air as an illustration. The live demonstration and its noisy aftermath were not appreciated.

Although several such pranks dotted their student days, Fermi and Rasetti were viewed with trust by the faculty. They obviously were talented students, and in the third year of their studies, they, along with their contemporary Nello Carrara, were given the keys to the university physics laboratories. The trio had free range to explore and to see what could be accomplished with the equipment at hand. Excited at the prospect of doing research, they set to work bringing the equipment in the laboratory up to date and building what was needed.

Each of the three selected a different thesis topic. The thesis was to be divided into three parts: a general introduction to the subject, a description of the underlying conceptual questions, and finally, the results with an interpretation of a performed experiment. It would have been easy for Fermi to write a thesis based purely on investigations in theoretical physics, but a thesis had to conform to those dictates. Fortunately he enjoyed experimental work and the format posed no problem for him.

For his thesis Fermi chose an ambitious topic, the study of X-rays, or as they were often still called, Roentgen rays, the name honoring their discoverer. Efforts were made in Pisa years later to find the original copy of his thesis, presumably available in the university's library. The efforts proved futile until 1990, when the thesis was in fact discovered in the library. Fermi's had been filed erroneously under the name Terni.

Fermi was self-deprecating about his thesis. He wrote to Persico in the frank and intimate style the two adopted with one another, "I have a lot to do for my thesis, which I might add has turned out to be a solemn piece of filth." Nevertheless Fermi did publish two articles based on its results in *Il Nuovo Cimento*, the journal of the Italian Physical Society. But these were not his debut publications.

Fermi's first original work was on the subject of relativity, a topic very much in the news following the 1919 observation of a solar eclipse. During that event, in accordance with Einstein's theory of general relativity, the path of distant starlight was seen to bend as it passed close to the solar surface. It was rumored that only a half dozen people in the world could understand how this observation related to Einstein's theory. That was an exaggeration, but Fermi was almost certainly the only Italian physicist able to comprehend the details.

In contrast, a number of Italian mathematicians were current on the fine points of general relativity. Foremost among them was Tullio Levi-Civita, a recently appointed professor in Rome. He had done much of the early work in the Riemannian geometry that provided Einstein with the tools for formulating his theory, leading Einstein to quip that the two best things in Italy were "Spaghetti and Levi-Civita."

During the summer after his first year in Pisa, Fermi began to apply notions of general relativity to the effects of gravity on the motion of electrically charged particles. To do so he needed a new and better system of coordinates than the usual x and y axes. Location on a sphere such as Earth is specified by longitude and latitude. But what is the optimal coordinate system for specifying the paths through the curved space of general relativity? Fermi's solution, which now goes by the name of the Fermi coordinates, was a significant advance.

Fermi realized that mathematicians would be interested in this research but also that they were unlikely to read about it if were published in the *Nuovo Cimento*. It would be better to have his work in the *Rendiconti dell'Accademia dei Lincei* (Proceedings of the Academy of Lynxes). The Accademia accepted only articles presented by a member. Fortunately for Fermi, a recent Pisan member of the Accademia was happy to oblige. He brought Fermi's work to the attention of Levi-Civita and his colleagues at a January 1922 meeting. Shortly after, Fermi's paper was published. At the age of nineteen, Fermi was beginning to be known in Italian academic circles beyond Pisa's ancient walls.

5

· · ·

THE YOUNG PROTÉGÉ

After four years at the Scuola Normale, Fermi received a magna cum laude doctorate in physics in July 1922. The oral defense of his thesis was anticlimactic; some of the eleven examiners in black academic robes and square hats were repressing yawns. None of them shook hands with him or offered congratulations, as was the custom. For them, his presentation was too erudite.

Afterward, Fermi returned to Rome. In spite of his brilliance, he had no obvious prospect for employment. Lacking a mentor, he found himself stymied. A very real danger loomed that university-based Italian physicists would not recognize his contributions to the nascent field of theoretical physics, and that mathematicians would not regard him as one of their own. Who, then, would be his advocate?

The prescribed path for entering academic life in Italy was first a position as assistant to a professor and then a *libera docenza*, the qualifying title for being a teacher. After sufficient time passed, one entered a competition for a professorship. This meant presenting your publications to a panel of five professors chosen by the Ministry of Education, since universities were state institutions. The professors would make the appointment after scrutinizing the merits of all candidates.

Given the system, appointments often involved favoritism. In addition, even if lucky enough to become a professor, one initially was almost always assigned to a minor university. After a few years, one might be able to transfer to a major center such as Turin, Bologna, or Padua and eventually maybe even to Rome.

Fermi was lucky; an influential patron recognized his astonishing talent. Moreover, this man, Orso Mario Corbino, was connected politically as well as being extraordinarily astute. Born in 1876 in a small town on the eastern shore of Sicily, Corbino—in his own way—was almost as remarkable as Fermi. Corbino's father had a small spaghetti-making enterprise. His mother, though from a relatively well-to-do local family, had never learned to read or write, the norm for Sicilian women at the time.

Young Orso was sent to the nearby town of Catania for high school, entered university there, and then continued on to Palermo, the island's largest city. There he developed a passion for physics. Graduating at twenty, he taught high school for several years while continuing to conduct experiments. At twenty-eight he won a competition for a physics professorship at the University of Messina, Sicily's third largest city. Four years later, he was offered a professorship in Rome.

At the onset of World War I, Corbino shifted his research to topics benefiting the war effort. In doing so, he came into contact with economic, industrial, political, and military leaders. They all realized Corbino's organizational and administrative acumen as well as his technical proficiency. As he moved into their spheres, a broad recognition for Corbino's overall excellence followed. In 1920 he became a Senator of the Kingdom, a lifetime selection made by the king, and in 1921 he was named minister of public education.

These notable appointments, political and administrative, coexisted with the retention of his physics professorship. Though conscious of the honors, Corbino was nevertheless saddened to venture forth from the cloistered world of academia. Expressing this feeling in a 1922 speech delivered to the Senate, he lamented, "I became a Senator, I became a Minister . . . but I miss the world of science; above all, in the midst of the bitterness of politics, I regret having left behind the peaceful days spent performing experiments while surrounded by apparatus."

More than any other senior physicist in Italy, Corbino was aware of the extraordinary advances taking place in quantum physics and was distressed to see that nobody in Italy participated in them. Serendipitously, Fermi appeared in Corbino's office, unsure of how much time the illustrious senator would have for a new university graduate. This shrewd judge of talent detected the young man's promise and saw him as the answer to his dream of Italy as a serious contributor to modern physics.

Thus began a close relationship that would last until Corbino's early death from a heart attack in 1937. During these fifteen years, the older man would advise Fermi on both professional and personal matters while constantly smoothing the way for the growing and increasingly successful research group Fermi led. Though not a participant, Corbino took pride in the group's achievements and made sure he was aware of their progress on almost a daily basis.

But the first thing Corbino did for Fermi was to ensure for him a stay in a great research center in northern Europe. Sensing that his young protégé needed to be challenged, Corbino wanted Fermi to meet others who might be his equals. The Ministry of Public Education offered a yearly scholarship for study outside Italy to a recent university graduate in the sciences. Not surprisingly, the selection committee that included Corbino unanimously chose Fermi as its 1923 recipient.

Germany, then the world leader in science, was Fermi's destination in January of that year. Language difficulties were not a problem, for his grasp of German was excellent—though still more at a reading rather than conversational level. He even had written his childhood friend Persico an occasional letter in German, signing off as Heinrich Fermi.

Two schools of theoretical physics had emerged in Germany as training grounds for young physicists during the early 1920s, both concentrating on atomic physics. They were where Fermi was most likely to find his peers. One was in Göttingen. Its university had been a world center of mathematics for over a century, and now, with Max Born at the helm, it was also a world center of theoretical physics. Arnold Sommerfeld, whose *Atombau und Spektralinien* was the bible for atomic phenomena, had made Munich the second mecca.

Fermi decided to use his scholarship in Göttingen. Curiously enough, his stay there was neither especially happy nor productive. Although Fermi was not treated badly during his eight months in Göttingen, there is no indication of his having been perceived as especially promising or of his having formed any connection with Werner Heisenberg, his contemporary and a rising star there.

Fermi was surely made conscious of the low esteem in which German physicists held physics research in Italy. According to a close colleague, Fermi had felt the Germans "were very conscious of their capacity, of their preparation, of their ability. All others were coming to learn from them. And it was true. But they tried to make a point of this, to stress the point." This bothered the proud young man.

The lonely twenty-two-year-old Fermi wrote Persico with some irony about Göttingen, including a comical sketch of the German perception of atomic scattering and a portrait of a prototypical Göttingen woman physicist. Both were uncomplimentary. He assured Persico that given the woman's looks, there was no danger of his being summoned as best man for a wedding.

Both the foci and the ethos of Göttingen physics were unappealing to Fermi. He always pursued a physical picture rather than the mathematical formalism prevalent in Göttingen. In this respect, it's interesting to compare Fermi to three other budding theoretical physics geniuses who were his contemporaries. Unlike Fermi's, the talent of those three was immediately recognized in Göttingen. In addition to Heisenberg (b. 1901), there were Wolfgang Pauli (b. 1900) and Paul Dirac (b. 1902).

By 1930, all four prodigies had done Nobel Prize–caliber work, were established professors, and were attracting younger physicists from around the world to their respective centers in Leipzig, Zurich, Cambridge, and of course Rome. The four often worked on similar problems, and sometimes even on the same ones, but in markedly dissimilar ways. Each had a characteristic style, a reflection of his personal strengths and predilections. It may seem strange that style plays such an important role in theoretical physics, since science's results are often portrayed in an impersonal manner. But human passions and special abilities shape its achievements as much as they do in other endeavors.

Pauli and Heisenberg had been Sommerfeld's students together in Munich and in successive years were assistants of Born's in Göttingen. Dirac had been educated at Cambridge, hardly the physics backwater that Italy was. Unlike the other three, Fermi was self-taught. In addition, Fermi considered himself an experimentalist as well as a theorist, combining action with concepts.

The three others, in contrast to Fermi, were exclusively grounded in theory. Dirac wanted mathematical elegance and beauty to be his guide. He was known for his quirkiness, often brusquely answering questions with "Yes" or "No" or "That is not a question." Heisenberg almost failed his doctorate exam because he enraged the committee's experimental physicist by being unable to explain how a storage battery worked. As for Pauli, he prided himself on the so-called Pauli Effect, which said that a key piece of machinery would always break when he entered the room.

One cannot imagine any of these stories being told about Fermi. He operated easily in both the theoretical and experimental spheres. Years later, when asked how he had entered the experimental field, he laughed and said, "I could never learn to stay in bed late enough in the morning to be a theoretical physicist."

While effectively bridging experiment and theory, Fermi also had his limits. He would not make the intellectual leaps that Heisenberg became known for or formulate one of Dirac's aesthetic mathematical marvels. Nor was he as famously critical as Pauli. But nobody could grasp all the interconnected aspects of a problem and reach a conclusion the way he could, nobody was able to significantly probe as many areas of physics as he was, and nobody could estimate orders of magnitude of physical phenomena as surely and as quickly as he could.

All in all, Fermi did appreciate the contributions of the German community of theoretical physicists. As a pragmatist, he was cognizant that they were unlikely to follow the Italian scientific literature. Accordingly, Fermi adopted the procedure of publishing key articles in either German or English. For reasons of national pride, he usually submitted a parallel version to Italian journals.

Returning from Germany to Italy in late summer of 1923, Fermi found his interest turning increasingly to statistical mechanics, a subject that would allow him a deeper understanding of thermodynamics, the study of heat. Thermodynamics had been one of the great triumphs of nineteenth-century science, its laws a cornerstone of physics and chemistry. But since thermodynamics is confined to macroscopic quantities, a number of individuals in the second half of the nineteenth century began to seek its underpinnings in the microscopic objects that make up the macroscopic state. They asked such questions as "What does thermal equilibrium mean, and how is it achieved? What does temperature really measure, and how does disorder come about?"

The logic behind these questions can be applied to other arenas. Knowing the size of a city, the total number of its inhabitants, and their average age tells a great deal, but it is not sufficient for planning traffic patterns. A well-oiled machine may work perfectly, but it is only by understanding its components and how it is assembled that one can truly appreciate its functioning. Asking for analogous explanations from thermodynamics led physicists to questions of probability, a topic that would continue to interest Fermi throughout his career.

Fermi's fascination with thermodynamics and statistical mechanics was absorbing him, but he still had no job. Fortunately Corbino came to the rescue once again, arranging for him to teach a mathematics course for chemists and biologists at the University of Rome. At least he would receive a salary, even if not a generous one. Fermi was frugal and still living at home. His material needs were met, although he felt intellectually isolated. Other than his friend Persico, who had remained in Rome as Corbino's assistant, only mathematicians seemed to grasp the meaning of his research.

As in Pisa, there were many more mathematics chairs than physics ones in Rome. Four of the occupants had considerable international reputations. None was more distinguished than Vito Volterra, the most senior of the four, but Guido Castelnuovo, Federico Enriques, and Tullio Levi-Civita were not far behind. They welcomed Fermi enthusiastically, recognized his talents, and anticipated what he might contribute to advancing Italian physics.

Beyond being embraced in an academic circle, Fermi discovered that the mathematicians also formed a tight-knit social circle, and he was invited to be part of it. The group provided him for the first time with a community of peers. Their families and those of their close friends typically gathered together on Saturday night, usually at the Castelnuovos', for far-ranging discussions on everything from scientific advances to local gossip.

Other than being great mathematicians, Volterra, Castelnuovo, Enriques, and Levi-Civita had another common trait, one that would come to have a significant impact on Fermi's life. They were all Jews. This might seem a remarkable coincidence for a country that had only some forty thousand Jews, approximately a tenth of one percent of the total Italian population of forty million. It wasn't altogether accidental.

Ghetto walls had been torn down in the middle of the nineteenth century and Jews had finally been allowed full access to university life. Education had always been valued in Jewish culture, and young Jews stood out in a country where more than half of the children over the age of ten were still illiterate. Furthermore, mathematics, if only as an aid to becoming merchants, bankers, and doctors, had been a customary subject for Jews to study. They would frequently select mathematics or related fields once universities were open to them.

The beginning of the twentieth century was a time of great pride for Italian Jews. After centuries of being shuttered away in ghettos, they were fully integrated citizens of a new nation. By and large, they became extraordinarily patriotic. Large synagogues were built in Italy's major cities, replacing the hidden rooms in unmarked buildings where worship had previously been held. And when war was declared in 1915, Jews rushed to enlist. Volterra, then fifty-five years old, had enlisted as a lieutenant in the Army Corps of Engineers and set to work calculating artillery trajectories.

Sadly, those feelings of patriotism would erode in the 1920s and eventually turn to scorn as Jews saw their homeland reject them by aligning itself with Hitler's racist tenets. And Fermi's newfound colleagues in mathematics were among the first to feel the yoke of anti-Semitism.

6

THE SUMMER OF 1924

As spring approached in 1924, Fermi was worried about his mother's health. She had suffered from a number of pulmonary ailments and recently spent time in a sanatorium. By April, all attempts to save her were failing.

In anticipation of enjoying their retirement in a peaceful setting, Fermi's parents had purchased land a few miles northeast of Rome in a housing development intended for government employees. They were building a small dwelling there, but it wasn't expected to be ready until the fall of 1924. It was looking as if Fermi's mother would not live to see it completed. She died on May 8, having just turned fifty-three. Afterward Fermi seldom spoke of her, and when he did, it was mainly to praise her organizational skills. Her lingering depressions and favoritism for her deceased child had made her emotionally unavailable to him, more notable for her managerial and technical capabilities than her maternal ones.

Nevertheless Fermi felt the loss and that summer looked for consolation in the beauty of the mountains, specifically in the Dolomite peaks and valleys that lie north of Venice and south of the Austrian border. The spectacular scenery drew him to the region and he also found congenial company, since many of the distinguished Rome mathematicians vacationed there with their families. Eager to continue taking the brilliant young physicist under their wings, they were happy to have him join them. Fermi could discuss algebraic geometry with them and also take long hikes with their children, in their twenties, closer in age to him.

The death of Fermi's mother coincided with another death, one the whole country was focused on. It was the murder of Giacomo Matteotti, an overtly antifascist socialist deputy, which for many was a watershed in Italy's march toward totalitarianism. On the thirtieth of May 1924, Matteotti had delivered an impassioned speech in Parliament denouncing the fraud in recent elections and the intimidation of opponents by Fascist brigades. At its conclusion he had declared, "I have finished my speech, now prepare the speech for my funeral." Many feared his prediction would come true.

On the sixteenth of June, Matteotti's body was discovered in a shallow grave about twenty miles from Rome. He presumably had been stabbed while trying to escape after being kidnapped. It remained murky whether his murder had been ordered directly by Mussolini, but it was a consequence of the climate Mussolini and his fellow Fascists had created during their two years of controlling the country.

Despite the unrest sparked by Matteotti's murder, opposition to Mussolini was divided. The king was weak, ineffective, and not inclined to criticize the man he had named as his prime minister. Mussolini, emboldened, made his move on the third of January 1925. Appearing before the Chamber of Deputies, he challenged its members to impeach him, adding, "Italy wants tranquility and hardworking calmness. We will provide this with love, if possible, by force if it is necessary. You can be sure that the whole situation will be made clear within the next forty-eight hours." Applause broke out, a clarion call for the country to turn into a totalitarian regime. Mussolini began referring to himself as Il Duce, foreshadowing a similar move by a man to the north who became known as Der Führer.

Dissent was no longer tolerated. In November 1926, a party tribunal was constituted and a special police force was formed. The Organization for Vigilance and Repression of Anti-Fascism, or OVRA, with several thousand members, began to permeate all levels of society. It would become a model for Germany's Gestapo and Russia's NKVD, though fortunately for Italy it never achieved the levels of ruthlessness and efficiency of the latter two.

Fermi seemed oblivious to politics. His views were very close to those Rasetti held, described thus by the latter in a 1982 interview: "In the first few years, in 1922, Fascism didn't seem so bad. In fact a large class of Italians welcomed it, because the Communists were very powerful and disorganized all industrial production, disorganized the railway traffic. So at that time Mussolini seemed a fairly reasonable dictator. The first act that really disgusted the more reasonable people was the Matteotti murder, which happened in 1924." Fermi was among those "more reasonable people."

How had Italy, a nation with a healthy democratic tradition, come to this point? The country had suffered in World War I, and not only from loss of life. Profiteering during the war had angered many returning veterans; the difficulties they were encountering in finding employment had further exacerbated the situation. The warnings of a national disaster were in place. Inflation was rising, industrial strikes were breaking out, landowners were afraid of reforms that would threaten their agrarian holdings, the army was unhappy, and the king was indecisive. Taking advantage of a government that seemed powerless, Mussolini, both cunning and ruthless, inserted himself into the mix. With a flair for propaganda and bombastic oratory, he was above all an opportunist, willing to change his position in whatever way suited his ambitions.

The Fascist Party, established in 1919, initially gained little traction, but as discontent rose, the accompanying fear of a left-wing takeover grew stronger; Mussolini exploited the desire for what he would later call "tranquility and hardworking calmness." He organized squadrons of thugs armed with clubs, their aim supposedly that of keeping the peace. And because he was financed by the right wing, Mussolini's hold increased.

The situation was ripe for Mussolini's ascent. On the twenty-eighth of October 1922, he launched the famous March on Rome, directing black-shirted Fascist squadrons to advance on the capital. The army had been mobilized and could have stopped the marchers, but that might have led to a massacre. Everyone was asking whether the king would sign the order for the army to take action.

Fermi happened to be in Corbino's office on the morning of that march. Never having given much thought to politics, the twenty-one-year-old Fermi looked to his mentor for guidance. That evening he recounted to his family what Corbino said would happen if the king signed the order for action: "So many young men will die who were only in search of an ideal to worship and found none better than Fascism." When Fermi had asked what one might hope for if the king avoided the confrontation, Corbino had replied, "A hope? Of what? If the king doesn't sign, we are certainly going to have a Fascist dictatorship under Mussolini." Both scenarios were bleak.

The king decided not to sign the order. Instead, on the thirtieth of October he asked Mussolini to form a new cabinet. Three years later the takeover was complete and Fascism firmly implanted. Opponents of the regime were jailed or went into exile. The Gregorian calendar was abandoned in favor of dating years in Roman numerals, year I coinciding with the March on Rome. Mussolini proclaimed the inauguration of a new Roman empire.

Mussolini had proved himself a master in his use of propaganda. He did so in person, in the press, and in newsreels, which were becoming an increasingly popular means of communication. The slogans he mouthed were repeated incessantly. In late 1923, after Mussolini's utterance "Better a day as a lion than a hundred years as a sheep," a circus owner offered him a lion cub. Mussolini adopted it as a pet-in-residence. Naming the cub Italia, he made sure newsreels filmed him in his convertible Alfa Romeo driving in the nearby Villa Borghese Park, Italia in his arms. The symbolism of a wild beast tamed by its fearless owner was not lost on the populace.

The political situation gradually entered Fermi's consciousness, but in 1924 he ignored it. Thinking about career rather than politics, he was planning a three-month stay outside of Italy, supported by a new fellowship. In January 1923, John D. Rockefeller Jr. had founded the International Education Board for the purpose of "promotion and advance of education throughout the world." Part of its mission was to award fellowships to promising young scientists, allowing them to visit active research centers for periods of months to a year. Fermi was the first Italian to receive one.

He chose to use his fellowship in Leiden, the Netherlands, opting to do so because of Paul Ehrenfest, a Leiden physics professor with a lifelong interest in statistical mechanics as well as in quantum theory. In his early forties, the Viennese-born Ehrenfest, a wonderful lecturer and an astute critic, was a close friend of both Bohr and Einstein. He was also known for being personally warm and involved in the lives of his students as well as for a free-ranging intuitive style in research, one that contrasted with the more reserved approach characterizing many other senior theorists, especially those in Göttingen.

After reading a 1923 article by Fermi in the *Physikalische Zeitschrift*, Ehrenfest wrote him a letter telling him how impressed he was. Ehrenfest also told a young former student of his living in Rome to get in touch with Fermi. George Uhlenbeck, later a well-known theoretical physicist, arrived in Rome to serve as tutor to the Dutch ambassador's son. He and Fermi, almost exactly the same age, contrasted dramatically in appearance, Uhlenbeck almost a foot taller. But physics, not height, is what mattered. In the course of their meaningful exchanges, the young Dutchman sang the praises of both Ehrenfest and university life in Leiden.

The stay in the Netherlands proved valuable for Fermi. Fermi never voiced why his Leiden experience had been so much happier than his Göttingen one, but always spoke fondly of the Dutch group, their openness and their acceptance of people coming from a different tradition. He was admittedly somewhat taken aback initially by how unceremonious Ehrenfest was in his manner as well as by his sloppy attire. The Jewish mathematicians in Rome were always proper and formal; Fermi described Ehrenfest to Persico as "really very nice and wouldn't be out of place in a ghetto store for used clothing." Ehrenfest made it known that he considered Fermi extraordinary even when compared to the best young theoretical physicists of the day. And Ehrenfest knew them all.

The biggest surprise for Fermi was meeting Einstein, who was spending twenty days in Leiden visiting Ehrenfest. This forty-five-year-old man took a fancy to Fermi. Einstein, himself an outsider of sorts, engaged Fermi immediately. There was a meeting of the minds, a shared interest in quantum physics and statistical mechanics. The obviously pleased Fermi tried to make light of his meetings with Einstein and the interest shown in him by the world's greatest physicist. In the letter to Persico, he described Einstein as a "very nice person despite his wearing a wide-brimmed hat that gives him the air of a misunderstood genius. He has been taken by a great liking for me that he cannot help telling me about every time he meets me (pity that he's not a beautiful girl)." In this case, Fermi overcame his usual embarrassment about attention paid to him. His parenthesized notation may have alluded to his lack of success in finding a girlfriend who would accord him similar admiration.

Fermi's warm reception was not the only reason the Leiden stay was happier than the Göttingen one. Fermi had matured personally and intellectually in the two intervening years. Many papers in Fermi's *Collected Works*, published in 1962, include introductions by fellow physicists who knew him at the time each paper was written. Looking back as a sixty-year-old, Persico commented on a 1924 Fermi paper, saying it had "the characteristics of the more mature style of Fermi: a fundamental idea, at the same time simple and clever, is applied to several concrete problems of physical importance with the help of mathematical methods of sufficient approximation, but not better than warranted by the underlying physical hypotheses." No better description of Fermi's style in theoretical physics was ever written.

In that 1924 paper, "The Theory of the Collisions between Atoms and Electrically Charged Particles," Fermi demonstrated that if one knew the effect of electromagnetic radiation on an atom, one could also deduce the result of that atom's collision with a charged particle. The varying electric field the particle produced could be described analogously to how radiation was treated. The paper highlighted Fermi's interest in all aspects of physics, the formidable arsenal of tools at his fingertips, and his ease in moving from the very practical to the very formal.

Nevertheless, in spite of the growing recognition of Fermi's breadth, depth, and creativity, he still lacked an academic position in Italy. And he needed one.

7

FLORENCE

To obtain an academic appointment that matched his ambitions, Fermi briefly considered emigrating, but his attachment to Italy was such that he continued his attempts to advance in its university system. The milieu was not conducive since most Italian senior physics professors remained mired in nineteenth-century physics, unaware of or unwilling to accept the innovative ideas of relativity and quantum theory.

Fortunately a few Italian senior physicists were forward-looking and well positioned. Antonio Garbasso was one of them. Like Orso Corbino, he had succeeded in politics as well as in physics. He was both Florence's mayor and a physics professor. This meant, however, that he spent most of his time in his majestic Palazzo Vecchio mayoral office and not in a physics laboratory.

Garbasso and Corbino identified Fermi, Enrico Persico, and Franco Rasetti, all three in their early twenties, as the most likely carriers of the torch toward modernity in physics. After graduation, Persico was taken on as Corbino's assistant and Rasetti as Garbasso's. They now had to find a position for Fermi, recognized as the most promising of the three young physicists. There was an opening as a lecturer in

Florence. Although the position was neither prestigious nor well paid, Garbasso and Corbino thought it would suit Fermi until he could obtain a real professorship. Both urged him to accept the appointment, which would also reunite him with his friend Rasetti. Fermi agreed.

At the time, Italian universities had no campuses, so departments were housed in buildings throughout the host city. This meant Florence students had to scramble, since the physics and chemistry departments were miles apart. The physics institute was located, with symbolic if not practical intent, near the villa that Galileo had returned to after his confrontation with the Inquisition and where he spent the last decade of his life under virtual house arrest. Situated amid a beautiful grove of olive trees on the Arcetri hillside, a few miles from downtown, the institute commanded a stunning view of the city. This made up for its lack of amenities, the principal one being no heating: the building's wintertime average daily temperature was in the low forties Fahrenheit.

During his first two years in Florence as Garbasso's assistant, Rasetti lived in a small room adjacent to the institute. With a bed, a desk, and a sink, it was sufficient for his needs. The wife of the janitor who took care of the institute would cook him simple meals. But in 1924, Rasetti's father died, and his mother decided she wanted to be near her only child. She sold her Pisa house and bought an apartment in Florence large enough to also house her son.

The timing was propitious. On his arrival in Florence at the end of 1924, Fermi moved into Rasetti's room in Arcetri. The facilities would be as simple as those in his Pisa dormitory and the meals no better, but he was feeling cheerful after his stay in Leiden. Physics was stimulating and Fermi was brimming with ideas. Besides, Rasetti had survived living there; he would not be more demanding than his friend.

Though Fermi and Rasetti were maturing as physicists, they still relished wicked practical jokes like the ones they had indulged in during their student days in Pisa. Rasetti described one that took place at Arcetri. Having gathered about thirty geckos in the surrounding fields, he and Fermi released them in their little dining area just before the janitor's wife came in with their lunches. Mayhem ensued and the two physicists roared with laughter, somewhat muted as they saw their lunches spill on the floor.

Fermi and Rasetti's outdoor activities were also continuations of their Pisa ones: skiing, hiking, and playing tennis. Tennis remained a favored sport of Fermi's. Other players would comment that though his style was not particularly graceful, his tenacity often made him able to wear down more accomplished opponents.

The two young physicists saw little of Garbasso. He came to Arcetri from the Palazzo Vecchio only three times a week to deliver lectures. Rasetti's limited tasks and Fermi's light teaching responsibilities left the two friends with time to conduct whatever research they could manage with the limited equipment at hand and the tiny budget available to them. Fermi had been working exclusively on problems in theoretical physics for the more than two years since his Pisa graduation. The possibility of conducting experimental research again intrigued him, particularly if he could do so with Rasetti and if it would serve to advance his knowledge of how quantum physics related to atomic structure.

Always an adept experimentalist, Rasetti was pursuing ideas in spectroscopy, a field once considered a backwater of physics and primarily of interest to chemists. Beginning in the 1850s, scientists had observed that an element, when heated sufficiently, emitted electromagnetic radiation. After further analysis, the radiation was seen to be different from element to element, but in all cases it consisted of distinct frequencies. This made spectroscopy an extremely valuable tool for chemical identification. Indeed, several new elements of the periodic table, most notably helium, were first detected by these means. This seemed, however, to have very little to do with atomic structure. Bohr would later say that it had been like trying to understand a butterfly by studying the colors in its wings.

That changed with Ernest Rutherford's 1911 discovery that the atom was composed of a tiny central nucleus surrounded by electrons. Two years after that, Bohr introduced his planetary model of the atom in which electrons moved in orbits about the nucleus, much as the planets circle the sun. Instead of a gravitational force, the negatively charged electrons were held in orbit by their electric attraction to the positively charged nucleus.

A volcano in the landscape of quantum physics had been smoldering since 1900, when Max Planck, in what he would refer to as "an act of desperation," introduced the concept of quanta. In order to reconcile basic principles of thermodynamics with experimental data, he had concluded that the energy contained in electromagnetic radiation was both absorbed and emitted in minute packets. The energy of a single packet was proportional to the radiation's frequency. Planck called these packets quanta.

Bohr's model of the atom had quanta emitted when electrons jumped from an orbit with higher energy to one with lower energy. The quantum's energy necessarily was equal to the difference between the energies of the electron in its initial and its final state.

The novel idea introduced by Bohr was to extend the notions of quantum physics to electron orbits in hydrogen by having their radii follow a numerical sequence of n squared times R, where n was any integer greater than or equal to one and R was the average radius of the smallest orbit. Since the energies of each electron depended directly on their position in the atom, those energies also followed a sequence. The integer n that characterized a given electron's energy was referred to as its principal quantum number.

Distinct electron orbits explained why hydrogen frequencies obeyed mysterious arithmetic rules. Measuring those frequencies, the field known as spectroscopy, became the primary tool for studying atomic structure. The butterfly wings had been meaningful after all. Bohr's model of the atom was so spectacularly successful in explaining the frequencies emitted by heated hydrogen gas that physicists immediately believed it contained some deep truth. But hydrogen, with only one electron, is the simplest of all atoms; extensions of the theory to other elements were not as convincing. Despite a subsequent series of successes over the next decade, the Bohr model met a corresponding number of failures. It was becoming clear that some key concepts were missing.

Searching for clues in spectra became one of the central problems in physics. Elliptical rather than circular orbits were considered, corrections due to the theory of relativity were made, and changes in spectra by external electric or magnetic fields were analyzed. The hunt was on for a deeper understanding of the atom.

By the early 1920s, each electron in an atom was being assigned not one, but three quantum numbers. They were roughly thought of as corresponding respectively to the orbit's size, its ellipticity, and its orientation with respect to an external magnetic field.

In early 1925, Fermi and Rasetti set out to examine a facet of this much larger subject. Rasetti made a specific proposal to his friend: they should carry out a set of investigations focusing on the polarization of light emitted by mercury vapor under the influence of an alternating magnetic field. He would provide the spectroscopic expertise and Fermi would be in charge of building the electric circuits.

The two obtained results notable enough to warrant being published in the prestigious foreign journals *Nature* and *Zeitschrift für Physik*, but they didn't move the field significantly forward. However, they constituted "the first instance of an investigation of atomic spectra by means of radiofrequency fields," a technique that was to receive numerous applications many years later, when radio frequencies became a more frequent tool in spectroscopy. The research was also distinctive because once again Fermi was showing his unique skill as both a theorist and an experimentalist.

On the whole, Fermi found the Florence position satisfactory, but he was eager to move on, win a competition for a professorship, and earn a decent salary. A professorship of theoretical physics still did not exist in Rome, so Corbino, supported by the Rome mathematicians, set out to establish one. It would be Italy's first. Fermi knew of the backroom maneuvers, but since administrative matters were holding up the process, he decided in the fall of 1925 to enter the competition for a professorship in mathematical physics at the University of Cagliari.

He wrote to Persico in October that "given the uncertainties in Rome, I plan to compete because I think it is advisable to have a double-barreled rifle, even if I don't find the idea of winding up on the Islands very pleasing." "Islands" is a reference to Sardinia, where Cagliari is located. Sicily and Sardinia are often the first destination of young professors ascending the Italian academic ladder.

Volterra and Levi-Civita, two of the five professors on the Cagliari professorship selection committee, voted for him, but the other three

were all physicists with little sympathy or interest in the field's modern developments. The Cagliari position went to a man thirty years
older than Fermi and undoubtedly less deserving. The committee's
three elderly members had felt obligated to reward their seasoned colleague rather than the young upstart. Their choice was even less
defensible since while they were deliberating, Fermi was writing a
seminal paper that would be considered a breakthrough in the world
of physics.

The origins of the paper's focus dated to 1924, when Fermi had
been stymied by problems he had encountered while applying quantum ideas to notions of statistical mechanics. A year later, in 1925, after
reading Wolfgang Pauli's new *Zeitschrift für Physik* article on what
came to be called the Exclusion Principle, Fermi was inspired.

The Exclusion Principle, for which Pauli was awarded the Nobel
Prize in Physics in 1945, was an extraordinarily important advance in
quantum physics. It put order into all the data accumulated in atomic
spectra by answering the question of how many electrons could occupy
an orbit. The principle postulated that no two electrons in an atom
could have all their quantum numbers be identical. Fermi found almost
immediately an innovative application for Pauli's idea by extending it
beyond the confines of the atom to the larger systems encountered in
statistical mechanics. One could imagine a gas of electrons, or electrons moving freely in a metal. They, too, would have energies obeying
quantum rules and following the Pauli Principle.

The notions Fermi introduced in doing this turned out to be the
key to understanding a wide variety of disparate phenomena, from the
difference between electrical insulators and conductors to the stability
of white dwarf stars. With this paper, first published in Italy during the
spring of 1926 and almost immediately afterward in the *Zeitschrift für
Physik*, twenty-five-year-old Fermi entered into the company of the
world's elite physicists—the only Italian in that select circle. It also was
a way of proceeding that was characteristic of all of Fermi's theory
work: take a clear physics notion, understand it in a way others had not,
and apply it to one or more important physics problems.

Fermi's paper was quickly appreciated in Northern Europe's great
physics centers. Many noteworthy applications of it followed, among

which were Pauli's explanation of previously baffling aspects of magnetism and Sommerfeld's study of electric current flow in metals. Now even more afraid that his protégé might be lured away from Italy, Corbino redoubled his attempt to keep him. Under his steady prodding, a competition for a professorship in theoretical physics at the University of Rome was set for November 1926. Furthermore, two other Italian universities, Florence and Milan, also determined that such a position would be desirable. Three professorships in theoretical physics had been created in a single swoop.

Bringing Fermi to Rome was not without hurdles. The city's university had two professors of physics. Rome's second, Antonino Lo Surdo, was not in favor of Fermi joining its faculty. He viewed the young man's possible arrival as a challenge to his standing and did not embrace the new generation, although he would have benefited from the freshness of its ideas. In mid-1920s Rome, while Corbino was looking forward to what the young would contribute to a new Italian physics, Lo Surdo was looking backward to the physics of yesterday and attempting to maintain the old guard's entrenchment. Like other Italian physicists of that era, he refused to accept either modern developments or their proponents.

In some ways Lo Surdo was a conservative mirror image of the progressive Corbino. Born only four years apart, both were Sicilians and had taught in Messina. However, Lo Surdo was no match for Corbino, who, while maintaining collegial civility, easily outmaneuvered him. The competition for all three chairs in theoretical physics had Corbino and Garbasso on its adjudicating committee. As expected, Fermi placed first, which gave him the position in Rome. Persico, placing second, went to Florence. And, with Persico's position as Corbino's assistant vacant, Rasetti transferred from Florence to Rome.

With the leverage of his new professorship, Fermi anticipated playing a major part in the overdue transformation of Italian physics research and teaching. Interest in science was slowly on the rise in Italy. Unfortunately, while there was progress on one front, Fascism was taking a toll on another.

In 1923 Italy inaugurated the Consiglio Nazionale delle Ricerche, or simply the Consiglio. It was founded largely as a result of the post–World War I realization that a flourishing economy and modern armed forces would require a country to have a solid scientific research base. Germany's powerful umbrella organization, the Kaiser Wilhelm Gesellschaft, had been founded in 1911 and the United States' National Research Council five years later. Italy moved to catch up by instituting its own variant. This was an auspicious start, particularly because the mathematician Vito Volterra, a man known for good judgment and impeccable honesty, was the Consiglio's first president.

Volterra would not last long in that role. In the wake of the 1924 Matteotti murder, he had asserted his integrity and independence by joining twenty other senators in casting a vote of no confidence in Mussolini's rule. The consequences of that vote were soon felt. With political loyalty trumping other considerations, Volterra's influence waned. His position as head of the Consiglio was not renewed when it expired in 1926. More ominously, even Volterra's presidency of the prestigious and supposedly independent Accademia dei Lincei was allowed to lapse when it expired, also in 1926.

Like other victims of totalitarianism, science was targeted by the regime's heavy hand. Mussolini insisted on having a loyal party member replace Volterra as head of the Consiglio. Guglielmo Marconi, not a scientist but a distinguished inventor who had shared the 1909 Nobel Prize in Physics for his contribution to wireless telegraphy, fit the bill. This enthusiastic Fascist, who had joined the party immediately in the wake of the March on Rome, became the government face of Italian research in science. Marconi was, however, not an academic. This allowed Corbino to maintain his influence in university circles. Fermi, who continued to cocoon himself from politics, depended on him for guidance and also for protection.

While these political gyrations impacted Italian science and played out in the public domain, the greatest twentieth-century physics revolution—quantum mechanics—was on the brink of forever altering the scientific landscape.

8

· · ·

QUANTUM LEAPS

The first glimpse of a decisive resolution to the ongoing quandaries of quantum physics came in June 1925. In Göttingen, twenty-three-year-old Werner Heisenberg was struck with a severe attack of hay fever. He retreated to the grassless North Sea island of Helgoland. In his words, as he was searching for a new way to attack the problems engulfing atomic structure, "There was a moment in Helgoland in which the inspiration came to me . . . It was rather late at night. I laboriously did the calculations and they checked. I then went out to lie on a rock looking out at the sea, saw the Sun rise and was happy." Heisenberg had taken advantage of the white nights of Scandinavian summer and worked until greeting a new day.

Heisenberg had done away with electron orbits, replacing them with a set of abstract rules based on observable quantities in electron motion. Assisted by Göttingen's senior theorist Max Born and his student Pascual Jordan, these ideas were soon extended into the full-blown theory that came to be known as matrix mechanics.

Its impact was not immediate, since the theory was cast in a novel mathematical formalism that almost all physicists had difficulty grasping. On the twenty-third of September, Fermi wrote to Enrico Persico,

"My impression is that there hasn't been much progress in the past few months despite Heisenberg's formal results on the zoology of spectroscopic terms." Franco Rasetti remembered being told by Fermi, "Now I'm trying to see what Heisenberg is trying to say, but so far I don't understand it." Though Fermi was respectful of the Göttingen school's achievements, Heisenberg's paper seemed to confirm his earlier belief that its physicists were overly reliant on abstract mathematical techniques. Fermi wanted a clear physical picture of what they were saying.

At first young Paul Dirac did not understand Heisenberg's theory either, but he soon recognized its essence. In early November 1925, he submitted for publication a paper entitled "The Fundamental Equations of Quantum Mechanics." The Göttingen trio, unaware even of Dirac's existence, was stunned reading his paper: he had reached the same conclusions they had. As Born wrote in his memoirs, "This was—I remember well—one of the greatest surprises of my scientific life. For the name Dirac was completely unknown to me, the author appeared to be a youngster, yet everything was in its way perfect, admirable."

Youngsters were indeed leading the rise of physics. By the end of February 1926, four remarkable papers, each written by a relative unknown, had appeared in the previous twelve months. Pauli, Fermi, Heisenberg, and Dirac had all rocked the quantum world. At twenty-five, Pauli was the oldest of the four. It is no wonder that Germans began referring to theoretical physics as *Knabenphysik* (boys' physics).

However, the revolution was not entirely led by youth. In early January 1926, while on a ski vacation in the Swiss resort of Arosa, thirty-eight-year-old Erwin Schrödinger was busy with something other than schussing down the slopes or charming the mysterious mistress who had accompanied him.

In a paper he wrote immediately after returning from Arosa, Schrödinger reintroduced the electron orbits that Heisenberg had done away with, but he did so with a new way to visualize them. In Schrödinger's version of quantum mechanics, the motion of an electron within an atom was guided by a so-called wave function. That unlocked the secrets of the atom for him.

Two months later, Schrödinger brilliantly showed that his theory, to which the name wave mechanics was given, was mathematically equivalent to matrix mechanics. In other words, every step of one had a mathematical analogue in the other. This meant there was only one underlying theory, though it apparently could be cast in two totally different forms. And Schrödinger had strong feelings about which one of the two was preferable.

He expressed them in a footnote to "On the Relation of the Heisenberg-Born-Jordan Quantum Mechanics to Mine." In that footnote, Schrödinger wrote, "I was absolutely unaware of any genetic relationship with Heisenberg. I naturally knew about his theory, but because of the [to me] very difficult-appearing methods of transcendental algebra and because of the lack of *Anschaulichkeit* [clearness], I felt deterred by it, if not to say repelled." The word "repelled" is an indication of Schrödinger's adamant feelings.

Planck, Einstein, and most senior figures in theoretical physics agreed with Schrödinger's assessment of the two theories, as did Fermi. Rasetti remembered his friend's reaction to reading Schrödinger's papers, a sharp contrast to the struggle he had seen Fermi undergo with Heisenberg's matrix mechanics. Fermi "understood it and then he poured it into the brains of a few people around him."

Schrödinger's papers quickly led a number of physicists to find ways to apply the ideas and techniques he introduced. The results were startling and gratifying. The wizards worked their magic. Even the problems that had already been solved saw their answers recast and clarified by making use of Schrödinger's wave functions. Dirac's treatment of statistical mechanics, in another paper by this young virtuoso, was a case in point. He quickly arrived at and then extended the conclusions that Fermi had reached six months earlier.

Unaware of Fermi's earlier work, Dirac published an article on the subject in the *Proceedings of the Royal Society*. Since his contribution was far from negligible, the resulting "Fermi-Dirac" statistics carries both their names. However, the identical particles it describes are known only as fermions. A large assemblage of them was called a "Fermi sea" and its edges a "Fermi surface." For the first time in the scientific lexicon, Fermi's name was used.

Dirac must have felt disappointed to discover that the Göttingen trio had independently obtained his main results about quantum mechanics and that Fermi had been the first to derive the statistical mechanics that a "Fermi gas" obeyed. But Dirac's originality was unquestioned.

Heisenberg was frustrated at seeing wave mechanics favored over his matrix mechanics. But he sensed that some key ingredients were missing. Bohr, who perhaps had a deeper understanding of the problems in quantum theory than anyone else, agreed with him. More tremors were still to come.

To decipher what might still be lacking, Heisenberg moved to Copenhagen in the fall of 1926 to work with Bohr. They were undertaking a search for quantum mechanics' true meaning. Working intensely, they formulated two new notions in 1927. Bohr's complementarity principle, emphasizing the complementary nature of matter as both particle and wave, and Heisenberg's uncertainty principle, establishing a limit on the simultaneous measurement of complementary variables, formed the basis of what came to be called the Copenhagen Interpretation of Quantum Mechanics. Its formulation brought to an end the second phase of the earthshaking change to physics.

The concepts Bohr and Heisenberg proposed were presented to the physics community during a prestigious conference held in Brussels during the month of October. Solvay Conferences had been taking place approximately every three years since 1911. They were intended to be meetings in which a few dozen of the world's leading physicists would gather for a week to address a major current scientific topic. The selection for 1927 was "Electrons and Photons," but those present knew the real subject would be quantum mechanics, its concept revolutionizing physics' foundations.

Significantly, no Italian was among the elite at the 1927 Solvay Conference, none deemed worthy of being invited to such an august gathering. One Italian, Enrico Fermi, would attend the next one, held in 1930. His presence there underscored Italy's arrival on the international physics scene.

Another international physics conference was held in 1927, a month earlier than the Solvay Conference. The Volta Congress, unlike the Solvay Conference, was a one-time affair. The nominal occasion for the meeting was the hundredth anniversary of the death of Alessandro Volta, the discoverer of the electric battery. The unspoken reason was the Italian government—and Mussolini in particular—wanting to show the world that Italy belonged to science's elite.

The meeting took place in Como, Volta's birthplace. Located on the beautiful eponymous lake, it was a delightful setting and invitees were treated to elegant lodgings and memorable boat rides. The indefatigable Fermi even managed to take a few hikes beyond the handsomely terraced gardens and up the steep slopes surrounding the lake, reaching huts with glorious views of snowy Alps to the north and the Italian plains to the south.

To ensure that nothing was lacking in organization or trappings, Mussolini demanded that the Italian electricity companies, dependent as they were on government backing, provide massive funding for all aspects of the meeting. Corbino quietly but wisely gave an honest assessment of the proceedings, observing that "Italy should have exhibited more physics and less hospitality and that it should not deceive itself that sponsoring a conference was a substitute for scientific achievement."

The invitation list was impressive. More than a dozen of the sixty-one individuals participating had already received Nobel Prizes, and several others would win the prize in years to come. Many hesitated before accepting the invitation. Arnold Sommerfeld, the highly respected senior professor in Munich, wrote to two eminent colleagues who had liberal views similar to his own, "I have serious reservations about attending because I assume the Italians will not forgo the opportunity of making it political and trotting out Mussolini." After wavering, all three went to Como. However, Einstein refused, wanting no part of an occasion he sensed might be used to burnish Mussolini's image.

The Volta Congress was notable in another respect. Though World War I had been over for almost a decade, feelings still ran high in many quarters. Since the war there had been no large international physics meeting that gathered together representatives from all the warring countries. The Volta Congress was the first. The leading physicists of Germany, France, Italy, England, Austria, the Netherlands, Belgium,

Denmark, and the United States attended. In that sense it was a huge success. It was also a propaganda success for Italy. Physics was another matter.

Unlike the Solvay Conference, the Volta Congress had a broad agenda loosely tied to electrical and magnetic phenomena. This allowed for a very wide range of presentations, but the conference lacked the intense focus that made the subsequent Solvay meeting so famous. Besides Bohr's first presentation of the Copenhagen interpretation to an international audience, none of the other more than fifty papers delivered was memorable.

Sommerfeld's presentation, however, turned out to be influential for Italian physics because its main thrust was to emphasize the significance of Fermi's recent work in statistical mechanics as the key to understanding the phenomenon of electric conduction in metals. The irony of Italian physicists having one of theirs achieve legitimacy through a foreigner's approval was not lost on Rasetti: "Everybody began to realize that Fermi had achieved something very important. So that was really the revelation of Fermi in Italy. His reputation in Italy came back through Germany."

When the week of meetings was over, the conferees were taken to Rome. On September 19, 1927, in Mussolini's presence, they were officially welcomed from atop the Michelangelo-designed steps of the Capitoline Hill by Marconi, the current president of the Consiglio. The scene was unabashedly dramatic. Mussolini and Marconi appeared to reign from on high. Afterward, Il Duce accompanied them to a gathering at his residence, Villa Torlonia. He was delighted to have an occasion for strutting in front of an illustrious audience, bombastically proclaiming the past, present, and future greatness of the nation he led. Abhorring posturing and ostentatious displays, Fermi must have cringed.

Though Fermi could not have been pleased by the meeting's political overtones, he was gratified to see the regard it afforded him. One souvenir of the time in Como was a photograph of him, Heisenberg, and Pauli sitting together smiling, the lake in the background. The three, an Italian, a German, and an Austrian, held the future of physics in their hands.

They knew they had already produced a revolution in physics. What they did not know was that without understanding the Pauli Principle, quantum mechanics, and Fermi-Dirac statistics, the world would not have been able to produce semiconductors, transistors, computers, MRIs, lasers, and so many of the other inventions that shape our life. In a very real sense we live in a world they created.

9

ENRICO AND LAURA

After Fermi's appointment on November 7, 1926, as Rome's professor of theoretical physics, he returned to the city of his birth, moving into the house on the city outskirts that had been constructed as his mother was dying. His father and sister now lived there. Maria had recently acquired a position teaching Italian literature at the same Rome high school attended by all the Fermi siblings. Sadly, Alberto Fermi was showing signs of a serious illness that would soon take his life. Maria and Enrico took turns sitting up with their father at night, but to no avail. He died on May 7, 1927, almost three years to the day since his wife had passed away.

That summer Enrico retreated to the Dolomites, just as he had done after his mother's death. Once again he found long walks in the mountains to be restorative. And on this occasion his sorrow was mitigated by joy. Although there had been passing fancies before, this was the first time he found himself truly in love. The object of his affection was a beautiful nineteen-year-old Roman woman named Laura Capon whom he had met the summer before.

The Capons were one of Italy's assimilated Jewish families that had risen to positions of relative prominence in the wake of Italy's indepen-

dence. Like many other such families, they were basically nonobser-
vant and seldom went to synagogue, though usually they married
within the faith. Their closest friends tended to be other Jews.

Laura's father, fifty-four-year-old Augusto Capon, was a career
naval officer. After Italy's unification, the military and the academy
were two careers to which many Jews gravitated. Having distinguished
himself both by his intellect and his valor during World War I, Capon
had risen to be the head of naval intelligence and would soon be made
an admiral. Like many officers, he was also a fervent monarchist. With
a happy family life and four children, of whom Laura was the second
oldest, he was comfortably well off.

Laura had intended to spend the month of August 1926 with her
parents and three siblings in Chamonix, the resort on the French side
of Mont Blanc. These plans were disrupted by the shakiness of the
Italian economy. Worried that Italy might succumb to the kind of
inflation plaguing other European countries, Mussolini had placed
restrictions on the export of Italian currency. In effect this necessitated
the Capons' staying in Italy.

An alternative plan was quickly formulated: the Capons would
vacation in the Dolomites near their friends the Castelnuovos. Guido
Castelnuovo, one of the Rome mathematicians who had befriended
Fermi, was only a few years older than Capon. He was a fellow Jew and,
like Capon, a Venetian by birth. The two families were close, particularly with these similarities and children of the same age.

In late July the Capons arrived in Santa Cristina, a small town in
the glorious east–west valley known as the Val Gardena. Located about
twenty miles from the main road to the Brenner Pass, it had already
become one of Italy's most desirable summer and winter resorts, with
ample opportunities for hiking, climbing, and skiing. The scenery is
spectacular, with jagged peaks soaring over lush meadows and high
plateaus. The church spires of small villages compete with mountain
pinnacles to create a wonderland of beauty.

As soon as the Capons arrived, Laura went to see her good friend
Gina, the Castelnuovo daughter closest in age to her. As Laura later
recounted, Gina greeted her by saying, "We are going to have lots of
fun. Even Fermi has written my mother asking her to find a room for

him." When Laura inquired who Fermi was, Gina replied, "You must know him, I am sure. He is a brilliant physicist: the hope of Italian physics, as my father says." And that is how Laura and Enrico got together that summer.

Laura had met Enrico two years earlier, albeit fleetingly. The occasion was apparently not filed in the annals of ardor or in the annals of memorable encounters. She described the 1924 meeting as follows:

> He shook hands and gave me a friendly grin. You could call it nothing but a grin, for his lips were exceedingly thin and fleshless, and among his upper teeth a baby tooth too lingered on, conspicuous in its incongruity. But his eyes were cheerful and amused.

Those gray-blue eyes may have peered at her differently two years later, as did her brown warm ones toward him.

During the weeks that followed the young vacationers went on day hikes along the region's many inviting paths, pausing from time to time to admire the glorious views. The group typically included a mix of friends and siblings. Laura discovered that though the young physicist was conscious of his rising reputation, he was not pompous; it was fun to tease him and he took the ribbings gracefully. She did not think Enrico was especially handsome, but there was something she found compelling about him. It probably was not the Tyrolean jacket and loosely fitting knickerbockers he typically wore in the mountains.

Everybody, parents included, trusted him to make the best decisions about excursions. He would pick the day's route, make sure all the hikers had what they needed, and see to it that the youngest hikers were not too heavily burdened. Fermi's knapsack was always by far the heaviest. Acting the role of mountain guide, he invariably took the lead and watched for trouble spots, not an infrequent circumstance on the Dolomites' tricky scree slopes. When one came along, he helped whoever needed a hand.

After that summer, Laura was delighted to learn that Fermi had been appointed to a professorship in Rome. The two could continue to see each other in the environs of the busy city. When summer again came in 1927, Laura returned with her family to the Dolomites, the scene of her budding courtship with Enrico. He returned as well.

Fermi fit nicely into the Capon family mix. The Capons were upper-class, but their standing was based on achievement, not inherited wealth or rank. Fermi felt at ease with them and they, in turn, were not skeptical of Laura's choice of a suitor, although he was not Jewish and not raised in their socioeconomic bracket.

Though both Fermi and Laura had grown up in Rome, their circumstances differed considerably. The Capons lived in a house located not far from the Fermis', but their beautiful garden-enclosed villa had little in common with the apartments Enrico had inhabited. The Capons had servants to take care of household affairs, slept on ironed linen sheets, went on expensive vacations, and always spent a few weeks in the fall at Laura's aunt and uncle's imposing country residence in the hills above Florence.

At the end of the summer, Laura began her customary stay with her Florence relatives. It afforded her on this occasion a chance to quietly study for university exams later that fall, a time when many Italian university examinations were given. Laura had just finished her second year in Rome's university, having chosen general science as a major. Though the major did not involve an intensive study of physics, it did require attending Orso Corbino's introductory course and gave her at least some feeling for Fermi's vocation.

In early September, Fermi had departed for the Como conference that became so pivotal to his international recognition. Both of them sensed that their separation would not be for long. After more than a year of knowing each other, their romance had blossomed. Later that month, Laura remembered, she was disappointed upon learning that Fermi had purchased an automobile, since he had laughingly told friends he intended to do something crazy, either get married or buy a car. But just as with his decision to be either a theorist or an experimentalist, he soon did both.

The car, an egg-yolk-colored Bébé Peugeot two-seater convertible with a rumble seat, added a certain zest to Laura and Enrico's courtship as well as a degree of uncertainty, because the Bébé was unreliable.

Fermi always kept the hand crank by his seat to start the car and hesitated to take it for long trips. For Sunday excursions in the countryside, Franco Rasetti, who had a similar car, provided backup for automotive mishaps. It was fun for everyone.

Laura appreciated the closeness of the bond between Fermi and Rasetti though she could not help but notice their dissimilarities. While Fermi delighted in female company, Rasetti seemed not to care much about girls even though they were attracted to him. Laura commented that he examined girls with "dispassionate detachment, bending his head to one side for a better view, with narrowed eyes behind his glasses. He examined them, dissected them with his piercing look, as if they were rare butterflies or strange plants." Fermi's musings on the female species were more forthright. He had told Laura that he sought a wife who was blond, tall and strong, and from "country stock." Laura met no part of that profile.

Although close, Fermi and Rasetti diverged in other ways as well. Fermi was rapidly moving toward bourgeois values; Rasetti continued to be something of a loner, still living with his mother. For Fermi, the days of throwing cats in the air during a lecture or of setting geckos free to scare a cook were over. His penchant for pranks had changed into a kind of good-natured humor.

Undoubtedly Laura affected Fermi's demeanor and contributed to his general happiness. He much admired her wit, intelligence, and casual elegance. Nor was he insensitive to her beauty. Leona Marshall, a co-worker of Fermi's fifteen years later, remembered her own reaction to meeting Laura: "When I first met her in 1942, I thought her the most beautiful lady I had ever met." When Marshall commented to Fermi on his wife's beauty, she writes: "Enrico caught his breath and told me I could have no idea how beautiful the teen-age Laura was."

Laura had been seventeen years old when she and Enrico first briefly met. He was obviously smitten. Those feelings began to be reciprocated two years later, in the summer of 1926. Admittedly Fermi did not come from a prestigious background, but that was more than made up for by the brilliant future toward which he was obviously heading. As Gina Castelnuovo had first said to Laura, everybody knew he was "the hope of Italian physics." When Enrico proposed marriage, Laura accepted.

Although enamored of him, Laura found one trait of his annoying. It was her husband-to-be's insistence on repairing everything he thought needed repairing without asking for help from anybody else. Fermi ascribed the attribute to his mother. He told Laura about how his mother had contrived to fix a pressure cooker, creating her own version in the process. As he explained to Laura, "If she wanted something, she would make it for herself." Having observed this at an early age, the son followed in his mother's footsteps.

This trait emerged inconveniently on Enrico and Laura's wedding day. Fermi was late for their departure for City Hall, located on top of the Capitoline steps. When Laura nervously asked about what had delayed him, the groom told her that when he unpacked a new shirt, he discovered the sleeves were too long. Instead of quickly pinning them up, he had painstakingly sewn a fold in them.

A photo of the wedding party includes friends and family: Corbino and Rasetti were among the twenty-five stylishly attired witnesses attending the ceremony. On an oppressively hot day, women wore fashionable hats and flapper-style dresses. Laura, in a scalloped dress, stood elegantly next to her beloved, their arms locked, revealing Fermi's long white shirt covering his wrist. His efforts at shortening his sleeves had not been altogether successful. But the dominant figure in the photo is Laura's father, the admiral, whose height and dashing uniform set him apart from the others. He is clad from head to toe in white, from a jaunty officer's hat to spotless white shoes.

Fermi could not help thinking of when, less than a year earlier, he had been atop the Capitoline steps. At that time, he was among those greeted by Mussolini after the Como Congress, where he had been in effect anointed as a physics genius. His marriage to Laura was an anointing of another kind. The only cloud in the sky on that July day in 1928 was the dark cloud of Il Duce on those same Capitoline steps in September 1927. Little did the wedding party realize how the Fascist dictator would soon change their lives.

The ceremony had gone smoothly. There was only a civil wedding, since the Capons were secular Jews and Fermi a nonbeliever like his parents. Only Fermi's sister, Maria, a deeply pious Catholic, minded the lack of a religious rite. As soon as Laura and Enrico were declared man and wife, Fermi's best man, Corbino, came over to Laura, kissed her hand, and said, "Congratulations, Mrs. Fermi."

That afternoon Laura and Enrico boarded a two-engine seaplane that flew to Genoa. Commercial aviation had started in Italy only two years earlier, so this was an adventure. From there they boarded a train and were off for their honeymoon in an Alpine valley lying between the Matterhorn and Monte Rosa. Not quite twenty-seven, Fermi now had a wonderful wife, a professorship in Rome, a thriving career, and even a semifunctional car. It was remarkable to think that only six years earlier he had been an impoverished Pisa student with an indeterminate future.

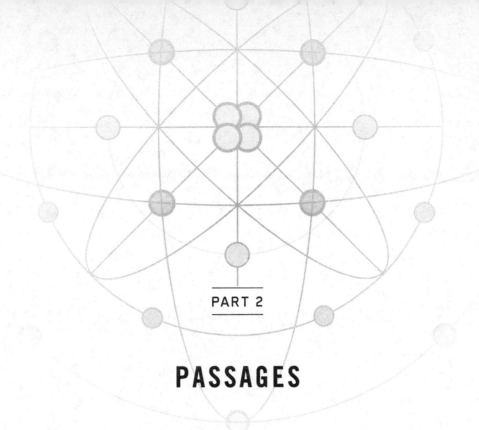

PART 2

PASSAGES

10

THE BOYS OF VIA PANISPERNA

The Colle Viminale (Viminal Hill), one of ancient Rome's proverbial seven hills, rises gently from the flat expanse linking the Forum to the Colosseum. Via Panisperna, one of the many streets that wind their way up the hill, housed the Rome physics department when Fermi became a professor there in 1927. The department, its address at number 89A, lay in the midst of the rapidly growing capital. Rome had almost quadrupled in population since 1870, the year of Italy's unification, a time when the city had only two hundred thousand inhabitants. The university had expanded accordingly, with most of the students locally based.

The villa at 89A Via Panisperna, with three floors and a basement, was spacious enough to accommodate the physics department. Orso Corbino and his family lived on the third floor. The second floor held a well-stocked library and the research laboratories of the department's three professors and their students. Classrooms and the shop occupied the first floor. The building had a serene atmosphere, in large part due to the palm trees and the bamboo thickets in the extensive garden surrounding it. A high wall shielded the plantings and villa from the noisy and dusty street and added to the semblance of an oasis of tranquility. As Emilio Segrè wrote, "I believe that everybody who ever worked there kept an affectionate regard for the old place and had poetic feelings about it."

Segrè, not a sentimental man, was clearly taken by this setting and all that it encompassed. At age twenty-two, he became part of an ambitious endeavor, nothing less than establishing Italian physics for the future. Fermi and Rasetti, both only twenty-six, had set the stage; they were building a unique research institute that would attract a top cadre of students. Segrè, a fourth-year engineering major at the university, would be the first to join them.

The recruitment of Segrè was not sheer happenstance; it came about because of mountain climbing. This was an avocation he shared, it turned out, with Rasetti. When Segrè heard that Rasetti had come to Rome, he contacted him immediately. The man quickly came to fascinate the younger climber. Aside from being a first-rate mountaineer, Rasetti spoke many languages, was widely read, knew everything about insects and plants, was doing serious research in a burgeoning field, and was the best friend and co-worker of the reportedly extraordinary Enrico Fermi. Listening to Rasetti talk made physics seem much more enticing than engineering. But Segrè was cautious. Was there really a future in Italy for physics?

An event in the summer of 1927 convinced him that there was. After having done a number of climbs together, Rasetti and Segrè set off in August for the Matterhorn, or the Cervino, as Italians call it. They ascended a ridge on the Italian side, known as a difficult route, and descended down the easier Hornli Ridge on the Swiss side. While on this expedition, Rasetti told Segrè he was planning to go afterward to a physics conference nearby at Lake Como. The conference, with many of the world's eminent physicists present, promised to be a festive occasion, one honoring the hundredth anniversary of the death of Alessandro Volta. Rasetti, still a youngster, had not been invited but he didn't anticipate problems in attending, at least the plenary sessions.

The thought of seeing all the great men of physics excited Segrè and he decided that "by tailing Rasetti, who in turn was tailing Fermi, I might be able to go to some of the lectures and see what was going on." For Segrè, "what was going on" became a turning point. He made up his mind to become a physicist, in no small part because of the prospect of an apprenticeship under Fermi and Rasetti.

Fermi was hoping to draw other promising researchers to Via Panisperna, but didn't quite know how to proceed. The indefatigable Orso Corbino helped him. During the previous spring he had paused in one of his lectures to second-year engineering students and a few general science students, including the future Mrs. Fermi. He announced that he was recruiting a brilliant young man named Enrico Fermi to the physics faculty in a few months. If any of them felt up to the challenge, Corbino was inviting them to participate in a field that was in the midst of a revolution. Great things were happening.

One of the students who felt ready to accept the challenge was Edoardo Amaldi, age eighteen. Two years earlier Amaldi's father, a well-regarded Rome mathematician, had taken the family for summer vacation to the Dolomites. There they joined the Castelnuovo circle of university mathematicians, including the Capons and Fermi. During that summer young Amaldi had gone on a number of hikes with Fermi and even on a long bicycle tour with him. Amaldi had been enormously attracted by the energy and joie de vivre of the new physics professor. Now, a little older and encouraged by Corbino, he was eager to shift from engineering to physics. The pattern of switching to physics that had begun with Rasetti after meeting Fermi was repeating itself.

These four, Fermi, Rasetti, Segrè, and Amaldi, formed the nucleus of what came to be known as the Boys of Via Panisperna. The Boys stuck to the same basic Italian schedule: working five days a week from early in the morning until one, taking a two-hour break for lunch at home, returning to work at three, and then staying until seven or eight in the evening.

There was one point of the workday that was sacrosanct. It occurred in the late afternoon, when the Boys would gather for an informal discourse by Fermi on a topic he chose or one that was perplexing the group. In his methodical way, Fermi would proceed, without consulting texts, to obtain all the formulas that might apply, solve the problem, and consider further questions. Fermi enjoyed this informal method of teaching and would continue to employ it throughout his life. His delivery was so smooth and seemingly effortless that students frequently did not realize how impromptu the session really was.

On Saturday mornings, plans were made for the next week's activities. Sunday was reserved for play: sometimes excursions into the Roman countryside, other times longer hikes in the nearby hills or trips to the beach at Ostia once the weather turned warm. Sunday outings were always group affairs for Fermi, but the members of the group varied according to who was in Rome and available for a daylong jaunt. In the early days, before he became a professor, the group might be his sister, Maria, Enrico Persico, and a few mutual friends. Later Laura, one or more of the Boys, and new friends might join. Arrangements were always flexible.

The many photographs of the changing group have a common look: smiling men dressed in casual clothes and similarly smiling young women in skirts and fashionable 1920s cloche hats, wearing hiking boots instead of elegant shoes. Conversations during the outings often had a cheerful teasing tone as they urged one another on as to who could be sillier. Occasionally the subject turned to physics, though never related directly to the research the Boys were engaged in. That was reserved for the rest of the week.

Playtime was not limited to Sunday day trips. During summer vacations, the Boys and their friends hiked in the Dolomites, and in the winter they skied in the Alps or the Dolomites, despite the scarcity of ski tows or lifts. Fermi, who loved all these activities, was an easy and sought-after companion. His piercing eyes shone with the intensity of his intellect, but his easy smile invited close friendships.

A merry mood and close camaraderie prevailed in Via Panisperna. One aspect of this was the bestowing of nicknames. This being Rome, several of them received ecclesiastical monikers. Fermi, regarded as infallible, was Il Papa (the Pope), and Rasetti was addressed as Cardinal Vicario (Cardinal Vicar), a nod to his position as Fermi's right-hand man. Segrè's judgmental disposition led to his being known as Basilisco (Basilisk), the legendary reptile capable of causing death with a single glance. Amaldi, with his youthful, cherubic, rosy face, was Fanciulletto (Young Boy). And then there was Corbino, whose ability to perform miracles, chiefly the raising of funds and the creation of assistantships for his young protégés, earned him the title of Padre-terno (God Almighty).

The Boys of Via Panisperna, originally a four-member group, soon grew to include an unusual fifth, Ettore Majorana. Majorana's brilliance had become obvious to his fellow students while he was studying for an engineering degree at the university. Segrè befriended the shy and self-deprecating young man and encouraged him to switch to physics. A short interview with Fermi convinced Majorana. A relative loner among the Boys, Majorana became known as Il Gran Inquisitore (the Grand Inquisitor) for his terse and critical manner, applied to the research of others, but particularly to his own work.

Il Gran Inquisitore rivaled Il Papa in the rapidity of his calculations. One day, at Fermi's request, Majorana had examined a complicated equation to determine how much energy it would take to ionize an atom, that is, to remove its electrons. Fermi was trying to approximate it numerically. Over the next few days Majorana retreated and then returned to Via Panisperna to compare his answers to Fermi's. To the astonishment of the Boys, Majorana had solved the equation analytically, his answer agreeing with Fermi's numerical conclusions. Until then, they had thought nobody could rival Enrico.

The paper Fermi wrote on this topic, applying statistical mechanics to atomic physics, was greeted with considerable approval. However, its acclaim was not as high as Fermi had expected. He soon learned that that was because almost identical results had been derived a year earlier by an Englishman named Llewellyn Thomas. Since Thomas's paper had been published in the *Proceedings of the Cambridge Philosophical Society*, a journal unavailable in Italy, Fermi had not seen it. And so, just as he had previously scooped Dirac, Fermi was now the one to be scooped.

Since the phenomenon of independent findings was common before the age of rapid communication, the discovery was credited to both men and hence goes by the name "Thomas-Fermi equation." Having been scooped on this one occasion had a relatively small impact on Fermi, whose international reputation was growing. His career beginnings in Rome had been off to a strong start. Soon there would be other reasons to recognize the Pope.

11

THE ROYAL ACADEMY

Though Fermi's main emphasis was on research and teaching, he knew that he should reach out beyond the confines of Via Panisperna if he hoped to promote the growth of physics in Italy. He became active in the government's National Research Council and served as an editorial consultant to the *Enciclopedia Treccani*, the Italian rival to the *Britannica*.

He also began writing for the *Periodico di Matematiche*, a journal aimed at keeping secondary school teachers abreast of recent developments in physics and mathematics. Fermi's early articles in the *Periodico* demonstrated his dazzling ability to explain what had been achieved and what still needed to be answered.

Because of Fermi's rising prominence, he felt responsible for keeping the Italian general public informed. Physicists, chemists, and other scientists were aware that new developments were taking place in atomic physics, but what bearing did they have on everyday life? In a long piece entitled *La Fisica Moderna* (Modern Physics) that Fermi wrote in 1930, he tried to provide an answer. He began with the rhetorical question "What practical consequences have been or might be derived from such a great increase in our knowledge of matter's intimate structure?"

His reply, still valid, is that it usually takes years, if not decades, for applications of fundamental new insights to be developed. But Fermi reassured readers "that the work of scientists is not distancing itself from life, losing itself in the pursuit of abstruse and purely abstract ideas." And he was right in that as well.

Fermi had an ulterior motive in accepting many of these writing commitments. They allowed him to earn money to supplement his relatively modest professor's salary, about ninety dollars a month. Though a man of simple tastes, he was aiming to have the kind of upper-middle-class way of life many of his colleagues enjoyed. A few years earlier, as a young bachelor, he had slept in a freezing room adjacent to the Florence Physics Institute; that was no longer acceptable for a married man with a growing professional reputation. Fermi now strove for a lifestyle that included a comfortable residence, interesting vacations, a housemaid, and occasional entertaining.

This was the mode of living that Laura had grown up with and expected to continue. Her dowry allowed them to purchase an apartment but they had no savings. Conservative by nature, they both also hoped to have a little nest egg if something unexpected and untoward were to occur. As Laura wrote, "Enrico felt that we needed more, not to lead an extravagant existence, but to acquire a sense of security and to be prepared for emergencies."

Fermi thought of a strategy for achieving greater security. He would, with Laura's help, write a physics textbook for Italian high schools. In 1928 he had published a book, *Introduzione alla Fisica Atomica* (Introduction to Atomic Physics), but it didn't make any money. A high school textbook might be different. There was no such text in Italian that Fermi thought adequate. With this book, students would learn physics in a new way, not having to consult obscure texts as he had been forced to do.

Laura and Enrico set to work on the textbook after their honeymoon, a trip that mixed romance and physics. The passionate husband could not stop sharing his infatuation with his trade; as Laura correctly deduced, "I was to learn physics, all there is to know about physics." By and large, Fermi dictated the book's contents to Laura. If she didn't understand what her husband was saying, she would interrupt.

This would often evoke the response "It's obvious," which would then lead Laura to reply that it wasn't obvious at all. Cooperation had its limits.

Working on the book mostly during vacations, Laura and Enrico adopted a program of writing six pages a day. This meant that the five-hundred-page text took almost two years to complete. Laura opined that it "was mediocre prose [but] it still served its purpose of bringing economic returns for many years." That income would not be needed after all. By the time the book was published, the Fermis had acquired a far more lucrative funding stream. Unexpectedly, it was thanks to Mussolini.

Since the mid-1920s, Il Duce had wanted to establish a prestigious Italian academy that would bring together prominent scholars and artists. Such an institution already existed: the Accademia dei Lincei. Originally founded in 1603, the Lincei had not taken its modern form until 1870, when Rome became the capital of Italy. The new Italian government had then provided the Accademia with a distinguished residence by purchasing the beautiful Palazzo Corsini on the banks of the Tiber. But the Lincei was too independent for Mussolini's taste. It took stands that were not always pleasing to Il Duce.

Mussolini wanted an academy whose actions on the cultural front would be in line with Fascist doctrines, one in which Italy's very recent past would be glorified. He also wanted to personally select its members. Accordingly, in January 1926, he announced the formation of the Royal Italian Academy. It would take him three years to assemble the funding for its support and for the annual awarding of four Mussolini Prizes.

Mussolini's choice for the Academy's location was pointed: Villa Farnesina was directly across the street from the Lincei's Palazzo Corsini. The villa, a grand early-sixteenth-century building, was a jewel with ground-floor frescoes by Raphael. Mussolini made sure the message to Italy was not lost: his Academy was superior to the Accademia. He provided generous financial support for its members: a university professor would more than double his salary. Artists were similarly compensated. To underscore the Academy's political connection to the regime, Mussolini scheduled its first meeting on the anniversary of the 1922 March on Rome, the event that marked his ascent to power.

In March 1929, Mussolini announced the Academy's first thirty members. The list included composers, artists, and playwrights. It had also appeared likely that a physicist would be appointed. Corbino was the natural choice, but the Academy's by-laws stipulated that if one was a senator, as Corbino was, he was not eligible for membership. Lo Surdo, the second Rome physics professor, thought he might be named, particularly because he was an ardent Fascist. But much to everyone's amazement, Fermi was chosen. Given the stipend that came with the nomination and the fact that it was a lifetime position, Fermi's financial worries were over.

Corbino had certainly been the force behind Fermi's selection, but Mussolini must have been pleased to see that Fermi did not hold any expressed antifascist views and was not even a member of the Accademia dei Lincei. Ironically, Fermi would have been a member except for a supposedly accidental lapse by Lo Surdo. Corbino, away on a trip to the United States, had asked him, in his absence, to nominate Fermi. Upon Corbino's return, Lo Surdo claimed to have forgotten. The truth was almost certainly that he, jealous of Fermi's success, had tried to keep him out of the Accademia. When Fermi was appointed to the Royal Academy, Corbino must have felt avenged.

In keeping with Fascism's bombastic style, the appointed Academicians were required to purchase an elaborate uniform to wear at official gatherings. Complete with cape, silver sword, and plumed hat, this outfit was Mussolini's creation, an image designed by Il Duce to impress onlookers and ensure that Academicians were conscious of the honor they had been granted. Unpretentious, Fermi found wearing the uniform embarrassing and went to some lengths to avoid being seen in it.

Though certain occasions could not be avoided, Fermi was probably the only Academician to arrive in a yellow Bébé Peugeot rather than a chauffeured limousine. An oft-told story revolves around his driving the Bébé to a high-powered government meeting. As Fermi—in his undistinguished car—approached guards blocking the access road, he asserted, "I am the driver to His Excellency Fermi. And His Excellency would be very annoyed if you didn't let me in." When Fermi recounted the story, he underscored that in both comments he had told the truth.

More disturbing to Fermi than the Academy's ostentatious trappings was the obtrusive hand of Fascism in the selection of its members. The rumor spread that Federigo Enriques had been in the initial group of thirty Academy members, but a lesser mathematician had been chosen at the last minute. Was Enriques's being Jewish the reason, or was it because the mathematician chosen was a Fascist? Government interference was also blatant in the conferring of the Mussolini Prize. In 1931, when Fermi nominated three distinguished mathematicians to choose from, all of whom happened to be Jewish, Il Duce summarily rejected each.

It is commonly said that there were no serious problems with anti-Semitism under Fascism until Mussolini came under Hitler's influence in the late 1930s. Mussolini was open about his longtime affair with Margherita Sarfatti, who was Jewish and an ardent promoter of artistic modernism. It does seem more than coincidental, however, that the Royal Academy never had a Jewish member.

Probably the insidious exclusion of Jews was partially due to a rapprochement between the regime and the Catholic Church. The schism dated back to 1870 when, in the final event of Italian unification, the Papal State was defeated and Rome captured. The Pope refused diplomatic recognition of the nascent Kingdom of Italy and, in turn, Italy did not recognize the Vatican State. The so-called Concordato or Lateran Pacts of February 1929 put an end to that almost-sixty-year hostile stalemate. The Pacts negotiated around this issue with a series of trade-offs. The Vatican was given a large sum of money and gained control over Italian marriage and divorce laws. The government, under Mussolini, recognized Catholicism as the official state religion and mandated religious education in public schools. Considerable benefits had accrued to the Catholic Church. But when Il Duce refused to kneel before the pontiff or kiss his hand, Pius XI began to see the high cost at which they had come. Once Italy was officially made into a Catholic country, Jews—by definition—were marginalized.

Two years later, Mussolini further tightened his control over Italy. In the fall of 1931, the government announced that all of Italy's university professors would be required to sign a loyalty oath professing allegiance and devotion to king, country, and the Fascist regime.

Again, this disproportionately affected Jews. Although Jews made up 0.1 percent of Italy's population, approximately 10 percent of university professors were Jewish. If one counted only those academics in the fields of science, mathematics, and medicine, the percentage of Jewish professors was much higher.

Of the more than twelve hundred and fifty professors at the time, only a dozen refused to sign the oath, among them the mathematician Vito Volterra. Some who took the oath justified their actions by reasoning that their places would just be filled by Fascist loyalists. Others maintained that it would be better to oppose the system from within. Many claimed the oath was a simple formality that meant nothing.

Fermi was not asked to take the oath, since he had already joined the Fascist Party a few days after his nomination to the Royal Academy. Although it was not a requirement for his appointment, it was an expectation. Fermi had been happy to oblige, since politics meant little to him. Physics is what mattered, and as long as he could pursue his research without undue interference, the rest did not concern him—an attitude shared by the other Boys as well.

Fermi's apolitical stance and desire to avoid clashes with the Fascist regime were well known. Enrico Persico, by now a professor of physics in Turin and nicknamed Prefetto di Propaganda Fede (Prefect of Propaganda for the Faith) by the Boys because of his success in spreading the quantum gospel, was aware of this. In a letter to Fermi recommending that he hire as his assistant Giancarlo Wick, a bright twenty-two-year-old Turin theorist, Persico alerted him to Wick's professed antifascism. Fermi's response was that he had no prejudices about political views but preferred having no public expressions of antifascism.

Wick, not a radical, was willing to curtail his political activities. He even took the oath of loyalty to the regime in 1937, when he became a professor. But in 1951, when the University of California regents asked all faculty members to swear they were not and had never been Communists, Wick resigned his Berkeley professorship rather than do so. Although he had in fact never been a Communist, Wick remarked to a friend, "I had once to take such an oath in Italy for mere survival reasons and I always regretted it." He never wanted to be in that position again.

12

CROSSING THE ATLANTIC

Ignoring the increasingly repressive political atmosphere in Rome, the Boys of Via Panisperna concentrated their energies on physics and were successfully making their mark. Edoardo Amaldi, Emilio Segrè, and Ettore Majorana had published—with a boost from Fermi—their respective first articles by 1928. They, as well as Rasetti, began to think of expanding their horizons. The prized Rockefeller Foundation fellowships, instituted in 1923, were becoming more widely available, and the Boys thought obtaining one might be a good vehicle for advancement.

Fermi had been the first Italian physicist to win a Rockefeller fellowship and had subsequently studied in Leiden. A few years later, Rasetti was awarded one. It would have been natural for Rasetti to go to one of the Northern European physics centers. Instead, always the most adventurous of the Boys, Rasetti was determined to see America, in particular the Wild West. He opted to spend the fellowship year at Pasadena's California Institute of Technology, familiarly known as Caltech.

The United States was not yet the physics powerhouse it would soon become thanks in large part to the efforts of a dynamic mix of individuals, some fully American-trained, such as Arthur Compton and Ernest Lawrence, and others who had studied in Europe, such as J. Robert Oppenheimer, Linus Pauling, and Isidor Rabi. That mix was further enriched by a number of European scientists who had emigrated. To them, America was a land of promise and a refuge from totalitarian regimes. It was to become home.

When Rasetti arrived at Caltech in 1928, he was impressed by the variety and vitality of American science. With his wide range of interests, he became friends with a broad spectrum of its biologists, geologists, and astronomers as well as with the physicists. On his return to Rome he regaled the other Boys with tales of his adventures and of strange customs, such as prohibitions against alcohol and the institution of faculty clubs where one mingled with other professors. He regarded America as a new world, one he urged them to see. Highlighting his affection for the United States, Rasetti imported a Ford Model A, undoubtedly one of the few Model A's in Rome and certainly the only one on Via Panisperna.

Rasetti's research work in California had been noteworthy and served as leverage for his obtaining an Italian professorship. Corbino once again held sway. To ensure that Rasetti was not separated from the other Boys, he managed to have a new physics professorship created in Rome. Rasetti filled it. Fermi began working immediately with him, as they had five years earlier in Florence. Again, Fermi was largely responsible for theory and Rasetti for experiment. This two-year collaboration was the springboard for a book Fermi published in 1934, *Molecole e Cristalli* (Molecules and Crystals). Translated into several languages, it became a standard text for both physicists and chemists seeking to understand the workings of the still new quantum mechanics.

As in Florence, Fermi also undertook independent research of deep theoretical significance while he was collaborating with Rasetti, who was focusing on more straightforward experimental projects. Quantum mechanics had shown how an electron interacts with an electric or magnetic field but not how it was able to either emit or absorb electromagnetic radiation. This was the question foremost in Fermi's mind.

A new subject, given the name quantum field theory, was in the process of being invented. Perhaps his own approach would help solve its mysteries.

Fermi was not the only one trying to advance the new subject. All the young theoretical physics geniuses, Dirac, Heisenberg, Pauli, and Jordan, were attacking the same problems. Fermi was less troubled than the others about conceptual difficulties. Instead, he laid out procedures to follow in an easily accessible fashion and then applied them to solving a number of relevant problems, with particular emphasis on explaining previously puzzling results.

Fermi's conclusions on quantum field theory were published in a number of shorter papers and finally summarized in a long pedagogical article. In addition to its methodology, it was notable in being Fermi's first publication in an American journal, *Reviews of Modern Physics*. The article's appearance in 1932 strongly influenced a whole generation of theoretical physicists, including Hans Bethe and Richard Feynman. Both future Nobel laureates considered the article pivotal in their own careers and in understanding field theory, commenting on its "enlightening simplicity." Feynman later wrote, "Almost my entire knowledge of QED (quantum electrodynamics) came from a simple paper by Fermi."

Fermi's decision to publish in an American journal was influenced by the very favorable impression of the country from his first visit. The trip came about because of an invitation by George Uhlenbeck, Fermi's friend from Rome and Leiden days. Uhlenbeck and his fellow student in Leiden, Sam Goudsmit, had just accepted faculty positions in physics at the University of Michigan.

Happily ensconced in Ann Arbor, they started a summer school for theoretical physics to bring American students together for a period of two months to hear lectures delivered by a few prominent physicists. Paul Ehrenfest, their Leiden mentor, had already agreed to come. Spurred by happy memories of his time in Leiden and encouraged by Rasetti's tales of America, Fermi accepted the offer. Laura joined him.

Neither Enrico nor Laura knew much about the United States. Laura admits she had never heard of the American Civil War and assumed Abraham Lincoln was Jewish, Abraham being a common name given to Jews in Italy. Neither of the Fermis knew English beyond basic communications: Fermi could read physics articles but little else. Their reactions to America diverged sharply. The country had little appeal to Laura, who counted the summer of 1930 as a failure in terms of any potential Americanization. In contrast, Fermi found he liked the United States very much. Within the university's ivory tower, Fermi was relatively impervious to societal changes and the effects of America's deepening economic depression. The spontaneity and lack of pretension of Americans is what spoke to him.

Laura, suffering from what she later admitted was snobbery, was less convinced. She, like many of her European friends, considered Americans uncultured and unrefined. While her husband had a busy life with other physicists, she had few friends and little to do. It is also likely that Laura was afflicted with morning sickness. The Fermis' first child, their daughter, Nella, was born on January 31, 1931, less than six months after their return to Italy.

Fermi returned to Ann Arbor in the summers of 1933 and 1935, catching on to spoken English, although his Italian accent would never disappear entirely. Laura remained in Italy both times. The new mother was reluctant to travel with her little one across the Atlantic, where she would sit in a strange house in Ann Arbor and not have the help of their tried and true nursemaid.

Returning to Rome, Fermi found that his rising fame as a theoretical physicist was beginning to attract visitors, most of them Rockefeller fellows from northern Europe, who spent appreciable amounts of time at Via Panisperna. Bethe was one of the first, arriving in February 1931. He was soon writing his Munich friend Rudolf Peierls, later Sir Rudolf, about the wonders of Fermi's approach to physics: "His ability to summarize any problem is amazing; he can immediately tell whether or not a paper makes sense . . . and his judgment of the theoretical and experimental (!) literature is infallible." Even after a short time in Rome, Bethe had recognized why Fermi was called the Pope.

Peierls, obviously intrigued, followed Bethe to Rome with his own Rockefeller fellowship a year later. Managing to extend his fellowship, Bethe joined Fermi again a year after that in the spring of 1932.

Via Panisperna had become a magnet for creativity and innovation in the field, much like Copenhagen under Bohr's tutelage, Munich under Sommerfeld's, Zurich under Pauli's, and Leipzig under Heisenberg's. Felix Bloch from Switzerland, George Placzek from Czechoslovakia, Edward Teller from Hungary, and numerous others came to Rome. They had all studied in more than one of those Northern European centers and sought to learn the less formal problem-oriented style advocated by Fermi. An added bonus was to spend time amid Rome's wonders.

Foreigners were not the only physicists coming to Via Panisperna. Word spread in Italy, and young Italian aspiring theoretical physicists such as Giulio Racah, Giancarlo Wick, and Ugo Fano also came to Rome. Largely thanks to Fermi they had become intrigued by the prospect of physics as a career path.

Pilgrimages to the Eternal City had been taking place for centuries. This was a new kind of pilgrimage, one whose endpoint was an encounter with a new kind of Pope.

13

BOMBARDING THE NUCLEUS

By 1930, Enrico Fermi was internationally recognized and Rome was becoming a world center of physics. To establish further prominence, it was timely to focus on the most exciting field in physics: the atom's nucleus. What transpired within the minuscule nucleus had largely been set aside, at least until the motion of the atom's electrons around its central core was understood. The problems raised were now coming to the fore.

In a talk Orso Corbino gave in September 1929, the Padreterno had said, "The study of the atomic nucleus is the true field for the physics of tomorrow," and he continued by asserting that to study physics "without an up-to-date knowledge of the results of theoretical physics and without huge laboratory facilities is the same as trying to win a modern battle without airplanes and without artillery."

Fermi followed up Corbino's envoi with a 1930 article in the *Periodico*, declaring that he, too, believed that studying the atomic nucleus was the foremost problem in the future of physics. He expanded Corbino's assessment by asserting that more than "up-to-date knowledge" would be asked of theoretical physics and that "we should expect that it will be necessary to modify the laws valid for the atom before obtaining a satisfactory theory of nuclear phenomena." The challenge was formidable.

The unexpected picture of the atom that Ernest Rutherford discovered in 1911 and Niels Bohr extended two years later did not attempt to explain the atom's nucleus. It did, however, seem plausible to interpret it as saying that the simplest of all nuclei, hydrogen, was nothing but a single particle with positive electric charge. The particle's mass, almost two thousand times as great as the electron's, accounted for why most of hydrogen's mass was concentrated in the nucleus. That positively charged particle was eventually given a simple name: proton.

But puzzles began right away. The next element in the periodic table of elements, helium, had a nucleus with two protons. But the helium nucleus's mass was approximately four times the size of a proton's. Something else had to be contained in it, but it was far from obvious what that might be. The same puzzle persisted all along the periodic table, with the imbalance between the number of protons and nuclear masses only growing.

Aside from the question of mass there was a glaring problem. What kept a nucleus together? The gravitational force was far too weak to make any difference. The only other known force, the electric one, explained how atoms were constituted: positively charged nuclei attracted negatively charged electrons. But rather than attracting protons to one another, electric forces caused repulsion.

Some other forces, unknown and powerful, had to be at work. As Fermi had written in one of his first publications, "These numbers show that the energy of the nuclear bonds is a few million times greater than those of the most energetic chemical bonds." But neither Fermi nor anybody else in the next ten years had been able to obtain a clue as to why those bonds were so strong.

That wasn't all. Even before he discovered the atomic nucleus, Rutherford had observed two kinds of decay in radioactive elements in which electrically charged particles were emitted, one negative and the other positive. In beta decay, an electron was emitted. In alpha decay, a positively charged particle was emitted. The alpha particle, far more massive than an electron, was ultimately identified as a helium nucleus. After the decay, the final nucleus differed from the initial one by having a different electric charge, plus one unit for beta decay and minus two units for alpha decay.

Since the existence of protons within the nucleus was well established, it was likely that two protons could unite inside a helium nucleus. But the presence of electrons in a nucleus's interior was dubious. The main attempt to circumvent this puzzle was to conjecture that a proton inside a nucleus had found an electron and bonded tightly to it.

This possibility had its pros and cons. It would explain how it was that electrons were observed to be exiting from a nucleus: they were contained in it to begin with. It would also explain the extra mass seen in nuclei other than hydrogen. Since the electron's and the proton's electric charges were opposite in sign but equal in magnitude, a bound-together electron-proton pair would have mass but no electric charge.

On the other hand, the conjecture had more than its share of problems, including one based on Heisenberg's uncertainty principle. But even putting that aside, how could an electron and a proton bind tightly together? And if they did, how could the electron free itself to form a beta ray? As if that wasn't enough, there was another problem: energy was apparently not conserved during beta decay. As expected, the difference in a nucleus's energy before and after it emitted an alpha particle matched the energy carried away by the exiting helium nucleus. The same was not true for beta decay.

All these interesting problems led Fermi in 1931 to organize an international weeklong meeting in Rome on nuclear physics to debate the issues. He was also hoping it would be an aid for him and the Boys to stay abreast of both experimental and theoretical developments in the field. Held at the Italian Royal Academy, it brought together leading experts from around the world. Underlining its importance and coincidentally the Academy's, Mussolini attended the inaugural meeting, held on October 11. The Italian press gave ample coverage to the event, not failing to note that the proceedings demonstrated "the depth and universality of Italian thought."

The gathering showed the Boys once again where the big challenges in nuclear physics lay, but it didn't provide them with a road map for how they could enter the field. Rome didn't even have equipment that was crucial for studying nuclear decays. That didn't deter Fermi. He and Amaldi enjoyed constructing experimental apparatus, so late in 1931 they set about building at least one of the needed items: a cloud chamber. This table-sized instrument contains vapor that records, via the formation of drops, the passage through it of electrically charged particles, alpha particles being prime examples.

Since the Via Panisperna Institute lacked what Amaldi and Fermi needed for building the chamber, they shopped in Rome in order to find just the right items, confounding clerks at the hardware stores with their strange assemblage of purchases. Fermi, a devotee of the do-it-yourself method, felt in his element by starting from scratch.

A cloud chamber is a delicate instrument. The one Amaldi and Fermi built could not compete with those built in more sophisticated laboratories. Their valiant attempt was, however, a useful lesson for them. It drove home to them the need for better experimental facilities and for further training in top foreign laboratories. Amaldi had already spent ten months in Leipzig and Segrè had left for Hamburg. Rasetti, the group's senior experimentalist, now traveled to the Berlin-Dahlem Institute, where Lise Meitner and Otto Hahn were doing state-of-the-art nuclear physics research.

As 1931 came to an end, a giant jump in the field of nuclear physics was beginning to occur. The key was to make use of an intense source of radioactivity to produce a beam of alpha particles and then use that beam to bombard an appropriate target. The technique was first developed by the German physicist Walter Bothe and then perfected by the French chemist Irène Curie and her physicist husband, Frédéric Joliot, in their Paris laboratory. They employed a target made out of beryllium, element number 4 of the periodic table, and searched for what sort of radiation was induced in the beryllium after it had been bombarded.

Exposing a layer of paraffin wax to the radiation that had been produced in the target, Curie and Joliot observed a copious production of protons. They concluded that quanta of electromagnetic radiation, particles known as photons, were causing protons to be expelled from the paraffin.

In Cambridge, Rutherford's right-hand man, James Chadwick, saw the note detailing the Paris experiment and reported its results to Rutherford. He was flabbergasted by his mentor's immediate response, "I don't believe it," a reaction Chadwick described as "entirely out of character." Rutherford and Chadwick had their own views on what caused those protons to be emitted: a particle whose existence they had been contemplating for years. It was massive and neutral: they had called it a neutron. When Chadwick now saw an opportunity to catch the evasive prey, he dropped all his ongoing research in order to do so. After two weeks of almost nonstop work, Chadwick proved that neutrons, not electromagnetic quanta, were responsible for knocking the protons out of the wax.

Curie and Joliot had misidentified what they had seen. In considering who should get the Nobel Prize for the discovery, Rutherford is reported to have said, "For the neutron to Chadwick alone; the Joliots are so clever that they soon will deserve it for something else." Three years later, in 1935, Chadwick was awarded the Nobel Prize in Physics for his discovery.

Although the neutron's discovery would eventually come to be seen as a watershed moment in nuclear physics and the true dawn of the subject, it took a while for it to become clear that was the case. Neutrons could account for the missing mass in nuclei, but there was a great deal of confusion about whether they should or should not be treated on a footing similar to protons. Were they elementary or were they composites? There was also no understanding of what kind of force could hold neutrons and protons together within the nucleus.

The first glimmers of hope in addressing this all-important issue came with three papers Heisenberg produced in late 1932 attempting to describe what such a force might be like. They provided a promising beginning on a subject that continues to this day. And Heisenberg was not the only one making such efforts; Majorana and others were as well. Chadwick and Rutherford had not been the only ones to have doubts about the Curie-Joliot report. In Italy, Majorana had shaken his head after reading it and had said to the other Boys, "They haven't understood anything. The effect is probably due to protons recoiling after being struck by a heavy neutral particle." None of them thought much about the remark, but a few weeks later they looked at Majorana with new respect.

Unfortunately, as was almost always the case with the hypercritical Majorana, he thought his own results were of no consequence. When Fermi asked Majorana if he might give at least a preliminary report on his work at a Paris meeting Fermi was planning to attend, the Gran Inquisitore had reportedly been furious, telling Fermi, "I forbid you to mention these things that are so stupid. I don't want you to go around discrediting me." Majorana was sadly showing signs of the paranoia and isolation that increasingly plagued him.

Fermi and the Boys tried to persuade Majorana to visit a few of the great European nuclear physics centers to gain some exposure to other theorists working in the field of nuclear physics and hopefully be convinced that his thoughts were not "so stupid." In particular, Fermi suggested a stay in Leipzig where Heisenberg was a professor. With a grant, Majorana left for Leipzig in January 1933. There, as Fermi had hoped, he found his work appreciated by Heisenberg, who even managed to persuade Majorana that his contribution to the theory of nuclear forces needed to be published.

Majorana's findings subsequently appeared in *Zeitschrift für Physik*, one of the last papers that foreigners sought to publish in this prestigious German science journal. The respect accorded to German science was being eroded by the diaspora of German scientists, many of whom were Jewish, fleeing Nazism's far-reaching tentacles. The research that was useful to National Socialism's racism was the distorted pseudoscience of eugenics. Basic science research continued, but soon that became reoriented toward weaponry and war.

14

DECAY

The Germany that Majorana found was very different from what it had been only a few months earlier. Adolf Hitler had become Chancellor on January 30, 1933. In an all-too-frequent scenario, political events were impacting the world of science. Laws were passed within two months giving Hitler total legislative and executive control; his dictatorship was firmly established. Another law was soon enacted that excluded Jews from civil service and restricted the number of Jewish students allowed in schools and universities. Göttingen's twin pillars of physics, the experimentalist James Franck and the theorist Max Born, went into exile.

Einstein didn't wait for the law that discriminated against Jews to be promulgated. After a prolonged visit to the United States, he had sailed back to Europe in March. When the ship docked in Antwerp, he went straight to the German embassy, renounced his German citizenship, and a few months later returned to the United States. He never went back to Europe.

Fermi, like most Italians, did not believe that the overt anti-Semitism sweeping Germany was a harbinger of what would soon be coming to his own country. Italy's Jews were a far smaller proportion of the population than Germany's and had risen to prominence in many positions of the government and the military after Italy's unification. By and large they considered anti-Semitism a thing of the past in their country. Laura's Jewish father, Admiral Capon, was not worried, nor was Laura. Fermi, too, put aside the recent developments in Germany and concentrated instead, as usual, on physics.

Fermi had benefited from the rise of Italian Fascism. His research group had been supported generously and his appointment to the Royal Academy had effectively doubled his university salary. Nonetheless, he was no fan of Fascism and also felt the intellectual void caused by the abrupt end of visits to Via Panisperna from the likes of Bethe, Bloch, Peierls, Placzek, and Teller. They were now searching for safe havens in the United States or the United Kingdom. Italy, with Mussolini firmly ensconced, was no longer appealing despite the excitement of physics there and the pleasures of exploring Rome.

Despite the deepening shadow of totalitarianism in Germany and Italy, the Solvay Conference was held as planned in October 1933, the seventh in the series typically held every three years in Brussels. As in the past, the Solvay Conferences convened a few dozen of the greats in physics to discuss a key topic. The choice in 1933 was "Structure and Properties of Atomic Nuclei." Most of the stars of physics, including Fermi, attended.

In discussions of new theories and experiments, the neutron and what it might mean for nuclear physics was uppermost, but another subject was often debated as well: beta decay. The problem of energy seemingly not being conserved when a nucleus emitted an electron continued to be troubling and there was no solution in sight other than abandoning energy conservation or accepting Pauli's problematic hypothesis of a mysterious undetected very light particle. While at the conference, a revolutionary idea of how to solve the decade-old problem of the missing energy in beta decay—one of the greatest mysteries in nuclear physics—began brewing in Fermi's mind.

The elegant simplicity of the theory Fermi proposed a few months later is wondrous. Seventy years after its formulation Marvin Goldberger, a post–World War II student of Fermi's, wrote, "Any physicist who has not read Fermi's 1934 paper on beta decay should rush out and do so immediately . . . it is the very epitome of what a scientific paper should be. The problem is stated clearly, a solution is presented, and the results compared with experiment. No smooth talk, no pretension, no promise that this is the first of a long series, etc. Just the facts!" Goldberger could have added that most of the last line of his description of Fermi's paper is also a good description of Fermi himself: "no smooth talk, no pretensions."

Amaldi and Segrè were the first to hear about the new theory. During the 1933 Christmas vacation they had gone skiing along with the Fermis and others in the Dolomites. This winter excursion amid the majesty of soaring mountains provided the setting for a jovial gathering of congenial friends relishing the fresh air of this snowy wonderland.

Amaldi's recent marriage to Ginestra Giovene gave the vacation an extra celebratory mood. Laura, in particular, was overjoyed to have Ginestra, only three years younger than she, become part of their circle. A close friendship between the two women complemented the one between their husbands. During the vacation, Ginestra realized she was pregnant, and the bonds between the women strengthened. Laura knew that she and Ginestra would soon also share the joys and complexities of motherhood. Years later, Amaldi would joke with his son Ugo, himself a physicist, that it had been an unbelievable vacation: finding out he was going to be a father and hearing about Fermi's theory.

After a hard day on the slopes, Fermi asked Amaldi and Segrè to come to his hotel room. Segrè, sore from falls on icy slopes, remembers thinking a hot bath might have been better than squeezing into a hotel room and listening to a new physics idea. Such thoughts vanished when Fermi said that what he was about to share with them was probably the best work he had ever done.

In his small room, Fermi laid out for them his new idea. Gravity and electromagnetism were not alone. A third force was operative, one that could transform a neutron inside the nucleus into a proton, an electron, and a very light neutron like the one that Pauli, the Viennese-born wunderkind, had tentatively proposed three years earlier. Fermi stated that if that transformation were to occur, the proton would remain behind inside the nucleus. On the other hand, the electron and

Pauli's neutron would escape immediately from the nuclear confines. This explained the mystery of how the electrons observed in beta decay could have resided, contrary to principles, within the nucleus. The answer was that they hadn't really been there. Created by the new force, they had been present only for an instant and then exited.

Nobody had ever developed a theory that could change a particle's identity, much less construct a theory with a third force. Fermi told his skiing friends that it was relatively easy to devise one that did that. What he didn't say was that it was easy for him, not for others.

Amaldi thought it was confusing to speak of Pauli's neutron and Chadwick's neutron. One had a mass that was at most a few times larger than the electron's and the other was as massive as the proton, two thousand times greater than the electron. He suggested a way, at least for Italians, to make it immediately clear that the two neutrons were really very different. Amaldi thought Chadwick's neutron should be referred to as *neutrone* (big neutral one) and Pauli's as *neutroncino* or more simply, *neutrino*. The first name has continued only in Italy but *neutrino* has become the universal way of referring to the little neutral one.

Fermi's theory was not immediately acclaimed. In 1934, *Nature*'s editors rejected the letter describing it that he submitted to them for publication. They felt "it contained abstract speculations too remote from physical reality to be of interest to the reader." Cited as a famous example of editors showing poor judgment, their response was not without its reasons. Two years earlier, physicists had been saying that the microscopic atomic world was made of electrons and protons held together by electric forces. Now they were being asked to take a leap of imagination into a world with new particles and new forces. They thought, as they said, that it was overly fanciful, "too remote from physical reality."

The editors were soon asked to accept for publication a letter hailing the correctness of that world. Bethe and Peierls, great admirers of the Pope's approach to physics, had written this new letter. In it, and similarly in one independently submitted to an Italian journal by Giancarlo Wick, they pointed out how Fermi's theory explained a recent sensational experiment Curie and Joliot had performed. Using the same type of alpha particle beam they had employed in the past to induce radioactivity, Curie and Joliot had observed nuclear decays in which positrons were produced.

The discovery of positrons, the antiparticle partners of electrons, had been the most unexpected and most surprising experimental result of 1932. Positrons had the same mass and were identical in every respect to electrons except for having the opposite electric charge, positive instead of negative. The theory of antimatter predicted that if an electron encountered a positron, the two would vanish, leaving only electromagnetic radiation (photons) in their place.

In their letter to *Nature*, Bethe and Peierls pointed out that the Curie-Joliot experiment offered convincing evidence for the correctness of Fermi's theory of beta decay. The positrons Curie and Joliot had observed had to be due to a proton inside an unstable nucleus being transformed into a neutron, a positron, and a neutrino. Before then it had still been possible to reject Fermi's theory. It was conceivable that a nucleus could harbor electrons, but as Bethe and Peierls wrote, "One can scarcely assume the existence of positive electrons inside the nucleus." If any had been present they would have quickly sensed the presence of a nearby electron: the two would have annihilated each other.

Bethe and Peierls also calculated the probability that, according to Fermi's theory, a neutrino could produce an electron or a positron when colliding with a nucleus. They did this to see if the technique could be used to detect neutrinos. Their conclusion was "There is no practically possible way of observing the neutrino."

But science marches on and what was impossible in 1934 had become possible by 1956. Two decades later, neutrino beams became a standard feature of experiments at large accelerator laboratories. The detection of neutrinos from the core of the sun had also been reported.

Yet the most amazing observation of these fleeting particles was to come.

When a massive star collapses, an enormous number of neutrinos are emitted in a few seconds and a supernova appears in the star's place. By the mid-1980s, three underground laboratories capable of detecting neutrinos from stellar collapse had been constructed—one in the

United States, another in Japan, and a third in the Soviet Union. There was little hope for such a sighting, since the last supernova formation close enough to Earth for neutrino detection had occurred hundreds of years earlier. Luck intervened. A ten-second neutrino burst, followed hours later by the appearance of a supernova, was observed on February 24, 1987, at 7:35 a.m. Universal Time by all three laboratories. It had taken those neutrinos 170,000 years to reach Earth. The unlikely, but possible, had occurred.

15

THE NEUTRON COMES TO ROME

Fermi's explanation of beta decay remains arguably his greatest contribution to theoretical physics. Its design was entirely his: nobody had considered anything like it and the idea was of enormous import to the world of physics. The possibility of a particle changing its identity would become central. So would the idea of forces beyond electromagnetism and gravity. Several other major contributions had established him as one of the world's most versatile physicists, but credit for his discoveries had usually been shared with others working on the same problem. Beta decay was Fermi's alone.

Despite the acclaim his theory received, Fermi never regarded his work on beta decay as other than an interlude. His goal did not veer from the one he had already set: to establish a solid and innovative experimental nuclear physics program in Rome.

In 1931, when he and Amaldi had tried establishing such a program, they had been hampered by the lack of equipment and by Via Panisperna's inadequate shop facilities. The failure had made them well aware of the need for improvement on this count. A year later, on September 30, 1932, Fermi wrote Segrè, "The problem of equipping the

Institute for nuclear work is becoming ever more urgent if we do not want to be reduced to a state of intellectual torpor."

By 1934, the Boys had more experience and more training, Rasetti's stay at Lise Meitner's Berlin laboratory proving especially valuable. They also had funds for farming out the construction of equipment. They were ready. The remaining question was what problem in nuclear physics they should attack. Inspiration came when they heard of Curie and Joliot's success in inducing radioactivity where it had previously been unknown.

Only a handful of radioactive elements had been identified before then, radium and polonium being the prime examples. The recent Curie and Joliot experiment showed that radioactive isotopes could be created. Natural radioactivity had been studied for over thirty years. Artificial or induced radioactivity was a total novelty. The structure of nuclei could now be studied in a way never before available; countless applications to medicine and biology were certain to follow.

Two years earlier, Curie and Joliot, erroneously interpreting an experiment, had lost the credit for detecting the neutron. This time they understood perfectly not only what they had measured, but to what it was due. Marie Curie, near death from radioactive poisoning due to her pioneering work with radioactive materials, was enormously pleased to see her daughter and son-in-law's achievement recognized by a Nobel Prize in Chemistry in 1935, the same year Chadwick was awarded the Nobel Prize in Physics for discovering the neutron. Rutherford had been right: they were "so clever." As someone who was a laureate in both physics and chemistry, Madame Curie understood the vagaries distinguishing each discipline and valued the contribution of each.

However, by the time Curie and Joliot received the Nobel Prize, they were following Fermi's lead. His ascent to prominence as an experimentalist had been precipitous. At the beginning of 1934, Fermi had been regarded almost exclusively as a theorist. By the end of the year he was well on his way to becoming one of the twentieth century's greatest experimentalists.

Fermi the experimentalist had made a major contribution by using neutrons rather than alpha particles to study induced radioactivity.

Success came very quickly with two crucial breakthroughs, the first in March and the second in October 1934. Recognition of them was immediate and was highlighted in his Nobel Prize citation four years later: "To Professor Enrico Fermi of Rome for his identification of new radioactive elements produced by neutron bombardment and his discovery, made in connection with this work, of nuclear reactions brought about by slow neutrons."

Using their beam of alpha particles, Joliot and Curie had not been able to induce radioactivity in any element past aluminum, number 13 in the periodic table of elements. Fermi, assisted by the Boys, was able to reach all the way to the table's end, uranium, number 92. And with slow neutrons he introduced a surprising and altogether new way of probing nuclear structure, one that would become extraordinarily important in future years.

Trying to use neutrons instead of alpha particles had seemed impossible. In order to obtain neutrons, one had to start with a beam of alpha particles and bombard a beryllium target with them: one neutron was produced for approximately every ten thousand alpha particles, a dauntingly small number. But Fermi realized he might have a chance to make good use of those comparatively few neutrons. They could travel straight to the heart of a nucleus, whereas positively charged alpha particles would be repelled by the nucleus's positive electric charge. Greater efficacy might compensate for smaller numbers; Fermi's estimate was correct.

Fermi was also lucky because the experiment he planned required a strong polonium source to produce the needed alpha particles, and he had one. This was not a foregone conclusion since such a source was both rare and expensive. Fortuitously, Corbino had introduced a bill in the Italian Senate in 1923 to create a special facility for handling and stockpiling radium and polonium; they were intended for use in research and medical therapeutic purposes. The Radium Institute was established in 1925 at Via Panisperna 29A and Corbino's assistant, Giulio Cesare Trabacchi, was chosen as its head.

Trabacchi, soon nicknamed Divina Provvidenza (Divine Providence) by the Boys, was an invaluable resource. His one-gram supply of radium had a market value of over a million Italian lire, an enormous sum at a time when university research budgets were generally only a few thousand lire. Its primary use was for medical treatments, but Divina Provvidenza would always supply the Boys with what they needed.

In early March 1934, Fermi began trying to induce radioactivity by using neutrons as projectiles. At first he and Rasetti worked together, but Rasetti soon departed for a prolonged vacation in Morocco, leaving Fermi to proceed on his own. The setup he adopted was simplicity itself. A small radioactive neutron source was introduced into a hollow cylinder coated with the material to be irradiated. After a time that varied from minutes to hours, according to needs, Fermi removed the source and ran down the hall with the cylinder to another room. He there inserted a thin Geiger counter into the cylinder to check if radioactivity had been induced in the cylinder's coating. Running down the hall was necessary to ensure that the counter was not measuring background radioactivity from the source.

Success came quickly. It was first reported in a one-page paper dated March 25, 1934, that Fermi wrote for *La Ricerca Scientifica*, an Italian journal often used for quick and short publications. This paper was the first of ten on the subject of induced radioactivity appearing in *La Ricerca* over the next fifteen months. They were pivotal in Fermi's career. His work over the next dozen years focused on neutrons. Beforehand, he had followed his whims, working on problems he found intriguing. He was now mining a rich vein that demanded his full attention. It was an unimaginable future. Eight years after *Ricerca I* was published, Fermi would prove that neutrons could initiate chain reactions, and three years after that he would be standing in the New Mexico desert watching what a chain reaction could yield.

Ricerca I announced to the physics world the creation of a vital and comparatively inexpensive way to study nuclear phenomena. *Ricerca II* delineated how radioactivity had been induced in elements of the periodic table all the way up to barium, number 56. This was more than four times as far along the table as Curie and Joliot's limit of aluminum, number 13. Appearing only two months after the French couple's groundbreaking paper, Fermi's achievement was sensational.

Fermi anticipated that groups around the world would rapidly follow his lead. If Rome was to play a major role, he had to work quickly and he would need help. Fermi began by asking Amaldi and Segrè to join him on the experiment. They agreed at once: the Pope had spoken. He then sent a telegram to Rasetti in Morocco. The Cardinal (Rasetti), summoned back to Rome, returned immediately. The Boys also needed an expert chemist who specialized in radioactive materials. Trabacchi's assistant, Oscar D'Agostino, joined their team. It was now a five-man group.

While there was no strict division of labor, Fermi was clearly the commander in chief. He made the ultimate decision of whether a measurement was valid or not. D'Agostino took the lead in chemical analyses, Amaldi oversaw electronics, and Segrè was in charge of acquiring the materials that would be irradiated. Since additional funds were needed, Fermi applied to the Consiglio for a supplement of twenty thousand lire, approximately $1,000 at the time. It was immediately granted.

Physics experiments had traditionally been carried out by only one or two individuals working together. By contrast, Fermi asked the entire group to cooperate. Primary assignments were specified, but all participated in every stage of the labor. Having worked at first on his own, Fermi was the sole author of *I* and *II* in the sequence of the ten *Ricerca* papers, but all five members of the group were coauthors of the subsequent ones, their names appearing in alphabetical order. These were probably the first papers to appear in the physics literature with five authors.

Word about what Fermi had achieved spread rapidly. On April 23, only a month after *Ricerca I*'s publication, Rutherford sent Fermi a letter. The tone he adopted reflects his somewhat tongue-in-cheek opinion that only experimentalists did real work. It started with "I congratulate you on your successful escape from the sphere of theoretical physics" and ended by asking Fermi to send him future publications on the subject.

Rutherford was not the only one to want those publications: copies of each of them were subsequently sent on a routine basis to the most prominent scientists studying nuclear physics. As Columbia University's young star physicist Isidor Rabi reportedly joked, "Now we will all have to learn Italian."

The atmosphere of camaraderie and playfulness that characterized the Boys' interactions was heightened by their awareness of working together on a potentially paradigm-shifting project in nuclear
physics. The waggish mood relieved the strain of making sure each
measurement was done correctly. The precise Cardinal, Rasetti, would
accuse the others of being clumsy. They would reciprocate by threatening to defrock the Cardinal. The Pope, who could be childish in wanting
to win every competition, would claim he was the speediest in carrying samples from the end of the second-floor corridor, where they were
irradiated, to the other end, where the Geiger counter was housed.
Amaldi would challenge him. The title of fastest remained contested.

Laura Fermi recounts the story of a distinguished Spanish visitor
coming to the Physics Institute looking for *Sua Eccellenza* Fermi.
Meeting Segrè on the ground floor, he was curtly told, "The Pope is
upstairs." After learning that *Sua Eccellenza* and the Pope were one
and the same, the visitor climbed the stairs, but he soon returned,
shaken. He had almost been knocked down by two young men in lab
coats running up and down the corridor. Further explanations revealed
to him who the two were: Fermi and Amaldi.

The first paper coauthored by all five members of the new Via Panisperna group concluded with a resounding message. The periodic
table of elements had to be changed. They had apparently succeeded in
producing number 93, an element with 93 protons. Up until then the
table had ranged from hydrogen, number 1, to uranium, number 92. It
now had a *transuranic*. The group was not altogether surprised.
Unlike the Curie and Joliot findings, the radioactive isotopes produced
by neutron bombardment typically decayed by electron emission.
Therefore, when an electron was seen being emitted after uranium had
been bombarded, the suspicion of having produced a transuranic was
natural. The subsequent check that the resultant nucleus did not lie
below uranium on the periodic table apparently confirmed the suspicion. The atomic number had increased by one unit, 92 turning into 93.

The Boys were, however, cautious in making their claim because of
the stir that finding the first transuranic would cause. Corbino had no
such hesitation. He wanted to let Italy know what the Boys had
accomplished. He soon found a way to do so. The academic year at the

Accademia dei Lincei was always brought to a close with a ceremony during which a member would give a formal address on a subject in his or her specialty. The honor in 1934 fell on Corbino; the topic he selected was "Results and Perspectives of Modern Physics."

On June 4, addressing his fellow Lincei members and King Victor Emmanuel III, who regularly attended the event, Corbino voiced his view of the new element, saying, "The investigation is so delicate that it justifies Fermi's prudent reserve and a continuation of the experiment before the announcement of the discovery. For what my own opinion on this matter is worth, and I have followed the investigations daily, I believe that the production of this new element is certain." The press, always on the lookout for a good story, found this one ideal.

With great fanfare, Italian newspapers trumpeted the discovery. The press, by now largely a tool of state propaganda, opined on how magnificently science flourished under Fascism. Some journalists even speculated that the new element should be named Mussolinium. One publication alleged that Fermi had given a vial of element 93 to the Queen of Italy. The international press struck a somewhat more skeptical tone, though the *New York Times* did print a two-column article under the banner headline ITALIAN PRODUCES 93RD ELEMENT BY BOMBARDING URANIUM.

Fermi did not share Corbino's certainty. He knew how easy it was to be wrong in making a science claim. Laura remembered being woken in the middle of the night by her normally sound-sleeping husband, torn about what to do. He felt his reputation was at stake. She advised him to speak to Corbino in the morning and see how they could address the situation. Subsequently a joint press release was issued, stating clearly that the announcement of a ninety-third element was preliminary. The disclaimer began with "The public is giving an incorrect interpretation . . . to Senator Corbino's speech" and ended with "the principal purpose of this research is not to produce a new element, but to study the general phenomenon." To counteract the overblown press coverage, Fermi also sent *Nature* a letter detailing the lingering uncertainties. At the very least, Fermi wanted to alert his science colleagues that he harbored doubts.

Aware that physicists everywhere were closely scrutinizing their publications, the Boys pursued their analyses at a feverish pace, check-ing and rechecking all their results. By mid-July they were ready to assert with greater certainty that they had observed element 93. Via Panispera, like other Rome laboratories, closed for vacation in August. The Fermis usually scheduled a summer vacation unless Enrico was teaching at the University of Michigan summer school or somewhere else. Fermi's penchant for working long hours was not always easy on Laura, and getting away was high on her agenda. The Dolomites, where they had met almost a decade earlier, certainly held a special place in their heart, but in August 1935 a stay there would have been less relax-ing because of considerable political tensions along the Italian-Austrian border.

Although there were strategic strains with Austria, relations between Italy and Germany were getting cozier. Two months earlier, on his first trip outside Germany or Austria since becoming Chancellor, Hitler had flown to Venice to meet with Mussolini. His German counterpart had not impressed Il Duce. Mussolini possessed, or so he thought, an imposing physique, whereas his visitor seemed pale and thin. Above all, Mussolini viewed himself as an established world leader dealing with an upstart aspiring to follow in his formidable footsteps. Hitler did indeed follow in those footsteps; six weeks after the visit he added the title of Führer, the German translation of Duce, to his designation as Chancellor.

The Fermis heard very little about this since they were en route to Argentina, Uruguay, and Brazil; interest in the new physics was high, and the large Italian colonies in South America desired to hear from a prominent citizen of their homeland. Fermi anticipated there would be lots to do in Rome during the fall, but he felt this excursion, almost two months long, would give him respite from work and be enjoyable for him and Laura.

They left their three-year-old daughter, Nella, in the care of a nurse-maid at the home of Laura's relatives, and the trip had the feel of a deluxe second honeymoon for Enrico and Laura. The Fermis were berthed on a luxury liner during the sixteen-day ocean voyage from Naples to Buenos Aires. Everything about the trip was first class, the ship and hotel accommodations, the dinners and welcoming parties. This was to be expected, because he was, after all, Sua Eccellenza Fermi.

On the return voyage the Fermis had as shipmates another Academician and his wife. Ottorino Respighi, the composer of *Fountains of Rome* and other symphonic pieces, tried to have Fermi share in the joys of music, and in return Fermi tried to explain physics to Respighi. Neither succeeded, but a friendship was struck.

16

THE RISE AND FALL OF THE BOYS

While Fermi was returning from South America, a small change had taken place on Via Panisperna. Bruno Pontecorvo, a young Pisan from a large well-to-do assimilated Jewish family, had been added to the group. Hearing of Rasetti's physics accomplishments, Pontecorvo had come to Rome to complete his thesis under the guidance of the fellow Pisan and family friend. Intelligent and gifted as an experimentalist, Pontecorvo became one of the Boys in the fall of 1934. As a mark of acceptance, he was given a nickname, Cucciolo (puppy dog).

Pontecorvo worked together with Amaldi in his first set of experiments as a member of the group. They derived a baffling result. Despite using the same target and the same source, the readings of induced radiation differed depending on whether the apparatus was placed on a wooden table or a marble one. Not understanding the inconsistency, Rasetti tried doing the experiments on his own. He obtained the same results Amaldi and Pontecorvo had. Everybody was now confused.

On the twenty-second of October, just back from a London conference, Fermi decided to see if he could discover the cause of the discrepancy. While making final preparations for the experiment, Fermi uncharacteristically hesitated, and then reached for a block of paraffin.

At the time paraffin wax was a typical supply in any well-stocked phys-ics laboratory. Without pausing, he put a two-inch-thick slice of the paraffin between the neutron source and the target. He saw at once that this yielded a dramatic effect on the induced radioactivity. The count-ing rate shot up unimaginably, by a factor of a hundred or more. The rise was simply astonishing. Other members of the Via Panisperna team, alerted by the commotion, swiftly congregated. Their first reaction was that the counters must be broken. They were quickly disabused.

Shortly before one o'clock, Fermi announced it was time for lunch. As was usual in those pre-cafeteria days, Italians went home for a meal with the family. Since Laura and Nella were still with Capon relatives in Tuscany, Fermi had a light lunch by himself. He was happy to be left alone, to think about the morning's experiment. At three o'clock, when the group reconvened, Fermi was ready with an explanation. But he could not help uttering, "What a stupid thing to have discovered this phenomenon without having known to foresee it." Fermi said the dif-ference in induced radioactivity was due to the neutrons in the beam being slowed down by hydrogen nuclei in the paraffin. When asked what ever possessed him to reach for the paraffin, Fermi gave them a response he often self-mockingly used when his interventions came out of the blue. He answered he had acted with *"intuito fenomenale"* (phenomenal intuition).

Fermi then told the group gathered around him that the neutrons probably had thousands of times less energy upon leaving the paraffin than they had upon entering. He continued his explanation, saying that less energy meant the neutrons were far more likely to be captured by the target's nuclei. More captures meant more induced radioactiv-ity. In addition, by moving slower, the neutrons would stay in the target longer, again increasing their chance of being captured.

Physicists had assumed that faster projectiles produced an increase in induced radioactivity. That was true for alpha particles. The oppo-site was true for neutrons. When slowed, they became more efficient at inducing radioactivity. Fermi told the other Boys this also explained the difference they had observed between doing the experiment on wood and marble tables. Neutrons that bounced off the wood table and reached the target had been slowed down. Those reflected by the marble had not. If Fermi's interpretation of how to increase radioactivity was correct, he added, water should have an effect similar to paraffin: they are both rich in hydrogen.

The simplest way to test this was to immerse the source and the target in a water tank. Although her account has been disputed, Laura maintained that the goldfish pond in back of the Institute was used to demonstrate this. She also added that the goldfish were unharmed, though perhaps disoriented by the strange presence in their waters. Whether the goldfish pond story is true or not, water's effect on neutrons was quickly confirmed.

The Boys recognized the significance of their finding: induced radioactivity had entered a new phase. That same evening they gathered at the Amaldis' house to write a report on their day's work. This was the natural meeting place because Ginestra had taken a position at La Ricerca, the journal in which they were planning to publish their result. If they wrote quickly, she could deliver it the following morning to the journal's offices. With Segrè acting as scribe, Fermi dictated the text; Ginestra's husband, Rasetti, and Pontecorvo paced around the Amaldis' living room shouting suggestions. Luckily the Amaldis' two-month-old baby seems to have slept through it all. Laura later heard from Ginestra that the Amaldis' maid, noting how exhilarated everybody was, had timidly inquired if they had been drinking.

The paper was written that evening. In a departure from the Boys' custom of listing authors in alphabetical order on their publications, Fermi's name was put first. It had been his discovery.

The technique discovered that October day would eventually also become a revenue source for the Boys. Upon hearing of the new discovery, Corbino told them it could radically accelerate the production of radioactive isotopes for medical purposes. He recommended they patent the procedure at once. They did so on October 26, 1934, four days after the first experiment. Amaldi, Fermi, Pontecorvo, Rasetti, and Segrè were listed as the five inventors. They agreed to divide any earnings from the patent seven ways, sharing them with Oscar D'Agostino and Giulio Trabacchi. Years later, after many hurdles, they would finally reap modest monetary rewards.

After the slow neutron surprise, the Boys' research proceeded at such a rapid pace that by the middle of February 1935, they had pre-pared for publication a lengthy paper describing fully their findings of the previous year. In a change from past patterns, Fermi submitted the forty-page paper to the *Proceedings of the Royal Society of London* instead of to German journals.

With a regime denouncing Jewish physics and expelling the likes of Einstein, Germany was no longer the physics magnet it had once been. Its publications were less vital, its research in decline, and its influence on the wane. Fully cognizant of this, Fermi was experiencing a greater affinity with the Anglo-Saxon world, particularly the United States, a country that was becoming a great power in physics and where he felt comfortable and appreciated.

The paper's eleventh section, the last and by far the longest, drew the most attention. It details, element by element, the Boys' systematic investigation of radioactivity induced by slow neutrons. The impressive list, a chemist's dream, is staggering in its thoroughness. A testament to how much work the Boys had done in just a few months, it impec-cably records slow neutrons' interaction with almost every element: hydrogen, lithium, beryllium, boron, carbon, nitrogen, oxygen, fluo-rine, sodium, magnesium, aluminum, silicon, phosphorus, sulfur, chlo-rine, potassium, calcium, vanadium, chromium, manganese, iron, cobalt, nickel, copper, zinc, gallium, arsenic, selenium, bromine, stron-tium, yttrium, zirconium, niobium, rhodium, palladium, silver, cad-mium, indium, tin, antimony, tellurium, iodine, barium, lanthanum, cerium, praseodymium, gadolinium, tantalum, tungsten, rhenium, irid-ium, platinum, gold, mercury, thallium, lead, bismuth, thorium, and finally uranium. Uranium was given the most exhaustive examination of all because of the importance of possibly having produced transuranics, elements 93 and beyond.

In section 11, the most telling paragraph comes near the end: it refers to research conducted in the Berlin laboratory of Otto Hahn and Lise Meitner. Aside from possibly Curie and Joliot, they were the world's foremost chemist-physicist team and close attention was always given to their findings. Hahn and Meitner had confirmed the Boys' observa-tions regarding transuranics. The Boys were relieved.

Years later, they all would be proved wrong. The German chemist

Ida Noddack had given what turned out to be the correct interpretation of the discovery of elements 93 and beyond. She suggested that the supposed transuranic might be the nucleus of an element that lay far down the periodic table, a fragment produced by the splitting of the uranium nucleus. Noddack criticized the Fermi group for basing their conclusions about transuranics on tests that had gone only a mere ten steps down the table from uranium.

The Boys dismissed her criticism, not even mentioning it in their paper. She had not conducted any analyses that might confirm her hypothesis. In addition, there was no physical theory in 1934 that would account for a nuclear splitting. There was also a touch of arrogance in the Boys' attitude toward her: Noddack was a chemist. They probably assumed she would not have understood how improbable such a split appeared to be from a physicist's point of view.

Hahn and Meitner also chose to ignore Noddack's paper. Admittedly they were only saying "probably due to transuranic elements," but that was also Fermi's outlook. The presumed existence of transuranics was soon endorsed by Irène Curie in Paris and further confirmed by experimenters in Berkeley, Ann Arbor, Vienna, Cambridge, and Zurich. Their findings having been checked by others, there was solid evidence that the Via Panisperna group had detected element 94 as well.

This prompted the Boys to contemplate what to call elements 93 and 94. The common tradition was for those who had discovered new elements to associate their names with the place where they had been first observed. Mussolini's advisers recommended *littorium* as a way of honoring the *littori*, a class of civil servants in ancient Rome who defended the supreme magistrate with *fasces littori*, weapons with thirty rods bound together and a protruding ax.

For their new element, the Boys did not want to adopt such a bellicose image, one frequently used as a symbol in Fascist propaganda. Corbino, with his keen sense of humor, told them that in any case the regime might not want to be linked to an element that decayed in minutes. Eventually, the Boys settled on Ausonium and Hesperium, Greek for Italy and for Land to the West.

The *Proceedings* paper, the apex of the Boys' research, was also their swan song, the last one they wrote together. Their accomplishments would continue, but largely with other individuals and in other places. After less than ten years, the Boys of Via Panisperna would become a thing of the past.

Part of this was due to the natural migrations of scientists. The group, after such an intense and productive run, was ready to disperse. But there is no doubt that the unraveling took place against the backdrop of a grave political climate.

In early 1935, Il Duce had begun planning a military campaign against Ethiopia, the land-locked independent African state. Since Ethiopia was a member of the League of Nations, founded after World War I explicitly to prevent future wars, the invasion put a severe strain on Italy's relations with its World War I allies, Britain and France. Mussolini was turning his back on Italy's historic partners and drawing closer to Germany, a country that had already withdrawn from the League.

The Italian invasion of Ethiopia, begun in early October 1935, drew further international condemnation when the commander of Italy's armed forces, Pietro Badoglio, ordered copious use of mustard gas. The Geneva Protocol had banned this weapon, an early example of science being used for corrupt purposes, after the devastation of World War I. Its renewed employment was widely met with opprobrium.

The sanctions imposed on Italy by the League were ineffective and victory in Ethiopia was declared on May 9, 1936. The League's powerlessness had not been lost on Germany or Spain. It was becoming painfully clear that they, along with Italy, seemed to be preparing for war.

In the summer of 1935, Rasetti had decided to take a year's leave of absence, opting to once again visit the United States. Segrè, recently married to a Jewish woman from Germany whom he met in Rome, also went to America that summer. Upset by the Italian political situation, he hoped to obtain a long-term position there. He did not succeed, but while in the United States he heard that he had been awarded a physics university professorship in Palermo, Sicily's largest city. He had entered the competition for the post as a fallback position in case nothing materialized abroad. Rather reluctantly, Segrè set off for Palermo. D'Agostino accepted a chemistry appointment at a government bureau located elsewhere in Rome. Pontecorvo departed for a research position at the Paris institute headed by Curie and Joliot. At age twenty-four, he wanted to see more of the world. By the fall of 1935, Fermi and Amaldi were the only Boys left in Rome.

It was hard still for them to consider moving abroad, especially since both of their wives were reluctant. Nonetheless, they found themselves increasingly unhappy about the direction Italy was taking. The physicists, bastions of rationality, could not fathom Italy's invasion of Ethiopia any more than they could justify Mussolini's call to fund the war by having women trade their gold wedding rings for steel rings or bracelets. With patriotic pride, Laura had joined the exchange ritual of Oro alla Patria (Gold for the Fatherland). These maneuvers were dismissed as propagandistic, only fleetingly interrupting the rhythms of domestic and work life.

Amaldi and Fermi were both ferocious workers. Amaldi recalled how the two "worked with incredible stubbornness. We would begin at eight o'clock in the morning and take measurements without a break until six or seven in the evening and often later . . . Having solved one problem we immediately attacked another without a break or feelings of uncertainty. 'Physics as *soma*' was the phrase we used to refer to our work performed while the situation in Italy grew more and more bleak, first as a result of the Ethiopian campaign and then as Italy took part in the Spanish Civil War." Amaldi borrowed the metaphor of soma from Aldous Huxley's *Brave New World* (1932), in which soma is a mythical drug of self-medication that eliminates feelings of stress and discomfort. Through physics as soma, the two researchers were able to block out the turmoil happening in the surrounding world.

The upshot of their year of labors between the summers of 1935 and 1936 is detailed in a fifty-page paper they published in *La Ricerca* at the end of May and republished in November 1936 in America's leading physics journal, *Physical Review*. Another paper, this time a mere forty pages, was also written by Fermi. It provided the theoretical physics underpinnings to experiments he and Amaldi conducted. Amaldi commented on its foresight: "The paper contains the seeds of nearly all of the important ideas on neutrons that Fermi developed in succeeding years."

It was an incredible effort and at the same time characteristic of Fermi. He was seemingly impervious to fatigue, frustration, or dissatisfaction.

Fermi's passion for physics flowered during 1934, 1935, and 1936, the years of his transformative accomplishments in neutron physics and beta decay. This had made those happy years for him despite the grim political climate.

In February 1936, another event added to those happy years. There was a new arrival: Laura gave birth to a baby boy. The tradition in Italian families was to name a son after his grandfather, but in this case Enrico wanted to honor the memory of his much-loved brother, who had died at age fourteen. The baby was given the name Giulio.

Fatherhood was again welcomed by Fermi, although, not uncharacteristically for those times, he was relatively uninvolved in the role. In her candid book, Laura has a chapter "How Not to Raise Children" that starts by describing how when Nella was born Enrico did not "dare to take his first born in his arms or even touch her. He looked at her from a distance with bewilderment and misgivings." By the time of Giulio's birth five years later, Enrico's behavior was only slightly less awkward. It was not that he did not love his children, but his devotion to physics—although not religious—had aspects of a higher calling: explaining the laws of nature.

17

TRANSITIONS

Toward the end of 1936, views of the future were looking bleaker and bleaker. The tightening vise of Fascism, the rumblings of anti-Semitism, and the drums of war affected Fermi's work. After a period of incredible creativity, it was as though the light of Fermi's genius was dimming. As noted by Emilio Segrè, "Fermi had developed a certain reticence." His commentary goes on to say that there seemed to be no special reasons for this. If one examines what was happening at the time, however, the reasons become discernible. The world as Fermi knew it was falling apart.

In January 1937, tragedy struck both personally and professionally. Following a brief bout of pneumonia, Orso Corbino died; he was only sixty-one. The Padreterno, Fermi's guide and protector, was gone. Fermi wrote a heartfelt eulogy of his mentor beginning with a description of their first meeting, Fermi a nervous twenty-year-old recent university graduate and Corbino an illustrious professor and a Senator of the Realm. The warm reception Fermi was given that day began an enduring friendship.

The personal loss Fermi felt was heightened by a bureaucratic decision to move the physics department to the new University City. The

Boys, or what was left of them in Rome, would no longer work in an inviting and intimate villa surrounded by palm trees, an island of calm surrounded by a bustling city life. The Boys also would no longer feel the embrace of the Corbino family, who lived one floor above their working space. Instead they were shuttled off to a complex erected at the city's edge, built in the mock-imperial style that came to be known as fascist architecture, ostentatious and seemingly grand on the outside but unwelcoming in its interior. The era of Via Panisperna had truly come to a close.

Luckily, Amaldi was chosen to fill Corbino's chair at the university. However, there was still trouble, because Lo Surdo, never a fan of Fermi and his group, was named to replace Corbino as director of the physics institute. Fermi had been the obvious successor for the position, but without Corbino watching out for him, political machinations triumphed. The Fascists could count on Lo Surdo. Fermi's prestige was such that Lo Surdo could do little to harm him, but he certainly would not go out of his way to help him and Amaldi.

The Rome physics upheaval in 1936 and 1937 was about more than Corbino's death and the move away from Via Panisperna. Major shifts were taking place in the field of nuclear physics, especially in respect to where the most vital research was being conducted and who was conducting it. Rome, struggling to keep up, was running the risk of obsolescence.

The glory days of Cambridge's Cavendish Laboratory, a leader in nuclear physics ever since the field's founding, were already over, its ingenious small-scale experiments a thing of the past. Lord Rutherford's death in October 1937 following a botched hernia operation was the coup de grâce. And the German centers in Munich, Göttingen, and Berlin, having to teach "Aryan" physics, were shadows of their former selves.

Nuclear physics was reaching a new level of sophistication. The old analogy of an electron's passage through a complex atom was discarded. Neutrons and protons were packed too tightly within the nucleus and interacted too strongly with one another for that picture to be useful. Niels Bohr and his Copenhagen associates, in part guided by the results of Amaldi and Fermi's 1935–36 experiments, were developing a new model of the nucleus. Making headway on the experimental front required new tools. Without them little progress would be achieved.

Fermi knew this. He had been a driving force in nuclear physics from 1934 to 1937, but his output in the next two years had little impact. It was unclear how productive he might be unless he moved to the United States, the country he repeatedly looked to as the likely future leader of physics. His visits there during the summers of 1930, 1933, 1935, 1936, and 1937 had made obvious to him the personal freedom America offered, and the work opportunities. A group of young men who had learned quantum mechanics in Europe were establishing schools of theoretical physics and chemistry in America, and they were being joined by a large number of refugees.

Talented and versatile young experimental physicists such as Luis Alvarez, Carl Anderson, and Edwin McMillan were also emerging in the United States. Then there was Fermi's contemporary, Ernest Lawrence, already famous for his development of the cyclotron. The praises of this machine's capabilities were being sung worldwide. Its beams of electrically charged particles were more intense than any other laboratory could achieve and the energies they reached were unprecedented. The consequence was that bombarding targets with the cyclotron's projectiles produced radioactive samples in previously unimagined abundance.

Fortunately for others, Lawrence was unusually generous with the riches spewing forth from Berkeley's cyclotron. Segrè remembers Fermi receiving a letter from Lawrence in the summer of 1935, asking him if he might find it useful to have a sample of radiosodium, a radioactive isotope of sodium. Lawrence suggested sending him a millicurie, an amount a thousand times greater than Fermi would have expected. After thanking Lawrence, Fermi pointed out that there must have been a mistake; Lawrence had surely intended a microcurie, not a millicurie. The reply was an envelope containing a millicurie of radiosodium. There had been no mistake.

But the Boys all knew that Lawrence's charity was not a substitute for having a cyclotron. Wanting to advance Rome's experimental capabilities, Amaldi, Fermi, and Rasetti spent the summer of 1936 in the

United States. All three came back to Italy in the fall enthusiastic about the prospect of building one in Rome.

Applying pressure to the Consiglio to fund the venture, Fermi wrote them in January 1937 that it would be "hopeless to think of an effective competitiveness with laboratories abroad, unless even in Italy a way is found to organize these researches on an adequate basis." Rasetti followed up that letter in an address to the Italian Society for the Advancement of Science. He pointed out to its members that the rest of the world had accelerators "that are functioning or in an advanced state of construction, twelve exemplars in the United States, one in France, two in England and one in Denmark."

The Consiglio's response was a grant of thirty thousand lire to develop a prototype of a cyclotron, a pittance compared to the million or more it would take to build a real one. Though the amount was not out of line with previous grants, it showed that Italy was unwilling or unable to provide the funds necessary for a major undertaking.

Still harboring hope, Fermi went back to the United States in the summer of 1937, this time to the West Coast. The cyclotron that had started functioning in Berkeley five years earlier was already being phased out and replaced by a larger and more powerful one. This progress was encouraging from the point of view of physics, but discouraging for what it presaged for Italy's future in the field.

While Fermi was in California, he received news that made chances of Italy having a cyclotron even more distant. Guglielmo Marconi, president of the Consiglio, had died of a sudden heart attack. The unexpected deaths of Corbino and Marconi left Fermi without his two special sources of support. Making matters even worse, Pietro Badoglio, the army general and former commander of the Italian troops in Ethiopia, was named as Marconi's replacement. The man who had ordered that mustard gas be used on the enemy was now in charge of Italian science.

One glimmer of light flickered amid the enveloping darkness in the fall of 1937. It made Fermi feel that perhaps what he had so carefully and successfully built to assure Italy's place as a center of exciting physics was not altogether lost. Without a real prospect of a cyclotron for experimental research, at least theoretical physics could prosper. The much-admired theoretician Ettore Majorana reappeared after a mysterious four-year withdrawal from physics. The man whom the Boys nicknamed the Gran Inquisitore had retreated to his Rome apartment in 1934, seldom emerging from its door. In essence, he had vanished.

Majorana's reentry was in the form of a brilliant article about neutrinos. Almost certainly Majorana's new publication was related to his unexpected declaration that he wished to be a candidate for a theoretical physics professorship. Professorships were scarce in Italy. Not since 1926 had there been a competition for these coveted posts. The announcement was heralded with a great deal of publicity. As was the norm in such cases, three candidates would be selected in priority order and three professorships awarded; in 1926 Fermi, Enrico Persico, and their friend Aldo Pontremoli had been the winners.

The 1937 appointment committee, which included Fermi and Persico, had essentially decided beforehand who the three would be, all deemed eminently suitable. That was before Majorana, clearly preeminent, announced his candidacy. Listing Majorana as first choice would mean dropping the third candidate, whose father was an extremely influential politician, called by Mussolini the "philosopher of Fascism." Denying this man's son the expected professorship might have ugly repercussions.

The answer, in one of those not altogether infrequent Italian accommodations, was to break from tradition. The appointment committee held a special session and managed to appoint all four candidates. Citing his exceptional merits, they awarded Majorana a special professorship in theoretical physics at the University of Naples.

Serendipitously, while Fermi's professional world was shifting, Laura's was settling. In 1936, she, along with Ginestra Amaldi, had published *Alchimia del Nostro Tempo* (Alchemy of Our Time), a popular treatise on atomic and nuclear physics. The two women had collaborated successfully on their favorably reviewed book. In 1937, they were enjoying its enthusiastic reception. This sealed a friendship that already was linked by physicist husbands, small children, and a love for Italy. Neither Laura nor Ginestra wanted to move from her homeland. Ginestra was sympathetic to Laura's reluctance to leave her frail, elderly father, who had retired from his post as an admiral in the Italian navy. He was alone after her mother's death in December 1935.

Yet Laura's fantasies about Italy somehow escaping fascism were rattled in the fall of 1937. Just weeks after Fermi's return from California, Il Duce undertook his first trip off Italian soil in more than a decade. Germany was his destination. Received with great pomp by the Führer, Mussolini was impressed by the sight of marching German soldiers, by a visit to the Krupp steel foundries, and by lavish receptions accorded him in both Munich and Berlin. Il Duce concluded his speech in the German capital by telling a crowd of close to a million that his and their country had to be united "in one single unshakable determination."

Mussolini stressed that commitment in December by having Italy follow Germany's lead in withdrawing from the League of Nations. This sent a clear signal to other nations that he and Hitler, already bound through their support of Franco's forces in Spain, would be acting together from now on. That unity was tested less than four months later by Germany's annexation of Austria, accomplished without Mussolini even being notified beforehand. Il Duce did not protest the deed: he obviously no longer felt the need for a buffer between his country and Hitler's.

18

STOCKHOLM CALLS

In hindsight it seems inevitable that Italy would follow Germany in enacting anti-Semitic legislation, but for the time being this trajectory did not appear to worry the Fermis. In January 1938, they moved into a new and larger dwelling near Rome's Villa Borghese Park. The Fermis had bought the apartment because, as Laura writes somewhat drolly, "I had been attracted by the idea of a green-marble-lined bathroom. It satisfied my ambitions of grandeur, which had been rising as Enrico's position had steadily grown better." She could now say, "I felt rich, well established, and firmly rooted in Rome."

These lines, written fifteen years after the events in question, reflect a much wiser woman thinking back about her naïveté at that time. It also indicates the degree of denial that permeated the Fermi household. Although there was some trepidation, neither Enrico, the child of civil servants, nor Laura, growing up in a loyal military family, was inclined to do anything other than support a government in power, even if they did not approve of its actions. Furthermore, given Fermi's status, there was no reason to believe his position would be threatened.

Fermi was shaken, however, in the spring of 1938, by both personal tragedy and political developments. On the twenty-fifth of March,

Majorana took the overnight ferry from Naples to Palermo. Before leaving he sent a suicide note to Antonio Carelli, the head of the Naples physics department. The next day Majorana wrote Carelli again, saying he had changed his mind and would be returning to Naples. He appears to have boarded the ferry, but that's where the trail ends.

All subsequent efforts to find traces of Majorana failed. The most plausible answer is that he committed suicide by jumping from the boat. But the body was never found and his family always insisted that no matter how desperate he had been, his strong religious beliefs would have precluded his taking such a step. The result has been a continuing debate in Italy, some suggesting that Majorana preferred anonymity, retired to a monastery, left Italy altogether, or even that he foresaw the advent of a nuclear bomb and escaped from having anything to do with it.

Assuming he had in fact committed suicide, his Via Panisperna companions mourned him. They regarded Il Gran Inquisitore's talents as sadly compromised by an overly critical personality that made him undervalue others' achievements and in particular his own.

The previous twelve months had seen many transitions in Fermi's life, all of them pointing to a darker professional and personal future in Italy. The moment was fast approaching when staying in their homeland would be untenable for the Fermis.

In May, Hitler and his entourage reciprocated Mussolini's visit to the Third Reich by traveling to what Il Duce was now referring to as the Second Roman Empire. Mussolini tried to match the might of the military prowess he had witnessed during his visit to Germany. The Italian navy mounted a grandiose display in the Gulf of Naples, ships leaving the harbor with airplanes flying overhead in formation. It was showy, but the visitors saw clearly that Italy's industrial output was in no way comparable to theirs. The Führer did, however, greatly admire the works in Florence's Uffizi Gallery, perhaps even thinking about the artworks he would collect after dominating Europe.

Germany's anti-Semitism had been very much alive even before Hitler came to power. There were few signs of such sentiments in Italy, generally regarded as the least anti-Semitic of all large European countries. Part of this may be due to the general tolerance exhibited by Italians, but it is also true that Italy's Jews, who numbered approximately fifty thousand, represented a far smaller percentage of the population than in most of those other countries, a tenth of what it was in Germany.

Mussolini's launching of an anti-Semitic campaign in July 1938 therefore took his compatriots by surprise. Many thought he was simply bowing to pressure by the Führer, but there is little evidence of this; it appears to have been entirely Il Duce's initiative.

His initial salvo came in the form of an order from the Ministry of the Interior: the Central Demography Office was to change its name to the Office of Demography and Race and to immediately take a census of all Jews in Italy. The *Manifesto degli Scienziati Razzisti* (Manifesto of the Racial Scientists) appeared in mid-July 1938. This document, claiming that a pure Italian race did not include Jews, was apparently mainly written by Mussolini. It was followed a month later by the first publication of a slanderous national biweekly, *Difesa della Razza* (Defense of the Race), a small but virulent anti-Semitic publication.

On September 1 and 2, the Italian Council of Ministers announced the expulsion of Jews teaching at any level of Italian schools, including of course universities. Jews who were already attending university would be allowed to complete their studies, but no new Jewish students were allowed to enroll. This was the tipping point for many.

The Fermis heard the news of the anti-Semitic edict while vacationing in the Dolomites. A heretofore reluctant Laura recognized that this was only the first of many more restrictions to come. She was also familiar with German racial laws that defined children with a Jewish mother as Jewish. Laura agreed, if only for the children's sake, to immigrate to the United States.

The first step in making the move was for Fermi to obtain a position at a major American university. On September 4 he wrote letters to four of the country's leading institutions, carefully mailing each of them from a different resort village lest the sight of all of them addressed to the United States from one small post office arouse suspicion. The decision to immigrate had finally been made, although the destination and other details remained to be decided. The Fermis returned to Rome in mid-September.

Early responses to Fermi's inquiries for a position were uniformly positive. Confident that further correspondence would result in at least one mutually acceptable offer, Laura and Enrico began making preparations to leave Italy in January 1939.

One of the things Fermi had never done, though Bohr had repeatedly invited him, was to attend one of the weeklong informal conferences at the famed Theoretical Physics Institute in Copenhagen. It was a ritual journey made by most nuclear physicists, and Fermi now viewed it as a fitting farewell to Europe. The decision was providential.

While there, Bohr broke rules of secrecy and divulged to Fermi that he was likely to receive the 1938 Nobel Prize in Physics. Bohr had a valid reason for alerting Fermi. Germany had prohibited its citizens from accepting a Nobel Prize of any sort as a protest for the 1936 Nobel Peace Prize having been awarded to the German dissident Carl von Ossietsky. Given Italy's and Germany's rapprochement, Bohr inquired if Italy would similarly force Fermi to reject a Nobel Prize. He assured Bohr that Mussolini had not yet adopted this policy.

On the way back from Copenhagen, Fermi stopped briefly in Belgium. From there, on October 21, he telegraphed his acceptance of the offer made by the head of Columbia University's department of physics. The next day he followed up with a letter entreating the chairman not to publicize his acceptance of a permanent professorship "for reasons easily understood." If the Mussolini government knew of his intent to leave the country permanently, they could throw up barriers to any passport requests.

Furthermore, he did not want to jeopardize the future of other physicists. It could do harm to Amaldi and the small group of scientists who had made Via Panisperna so vital. His thoughts were also fixed on his country's Jewish physicists. As he wrote to Columbia's chair, "Making use of my writing from Belgium, I want to ask about positions for some young Italian physicists who have been deprived of their positions because of the racial laws." The list of five was Fano, Pincherle, Racah, Rossi, and Segrè—all Jewish. It was an act of thoughtfulness and caring.

The offer from New York had been the most appealing to Fermi. Appreciative of their strong neutron physics experimental group, Fermi was familiar with the environs, having spent the summer there in 1936. He was also sure Laura would feel more at home in New York

City's cosmopolitan atmosphere than in a small city such as Ann Arbor, Michigan.

When Fermi returned from his Copenhagen/Belgium trip, he requested a six-month leave of absence from the University of Rome. His public stance was that he would be away at Columbia only for the spring term. The acceptance of a permanent professorship concurrent with a plan of long-term immigration was known only to Rasetti and Amaldi. Fermi also shared with them the prospect of a Nobel Prize.

Fermi had asked Columbia for a long-term visa for the family, one that would pave the road to permanent residency in the United States. A six-month American tourist visa was still relatively easy to obtain in 1938, although longer than that was difficult for non–northern Europeans. Quotas were not kind to Italians and other "southerners." Fermi, who prepared meticulously in all matters, did not want to leave anything to chance. Thanks to strong backing from Columbia University, the desired visa was granted.

With things falling into place, the expected phone call from Stockholm came at six in the evening on November 10. It was official: Fermi had won the Nobel Prize. The citation was read to him over the telephone. He was elated by the acknowledgment; in addition, the prize money would provide a handsome nest egg for settling in the new country.

It also meant a change of plans. Instead of leaving directly from Italy for New York, the Fermi family would leave a month earlier from England—after the Stockholm ceremonies in December. They would not return to Rome. That way they would not be required to change the prize money into lire and would be able to take the full amount, approximately $45,000, with them, thereby skirting Italian laws stipulating that not more than $50 could be taken out of the country.

Shortly afterward, having heard the news, the Amaldis and other friends came knocking at the door to lead an exuberant celebration. Ginestra insisted on a feast, which was duly organized and enjoyed. The joy felt by the small group was tempered by other news. The Fermis and their friends had heard it on the radio on that same day.

On November 11, in banner headlines, *Il Corriere della Sera*, Italy's leading newspaper, announced: THE LAWS FOR THE DEFENSE OF THE RACE APPROVED BY THE COUNCIL OF MINISTERS. New civil statutes defined who was a Jew and proclaimed that mixed marriages such as the one between Enrico and Laura were forbidden. Children who had resulted from such a marriage would be regarded as Jews unless they had been baptized before October 1, 1938.

Simultaneously, the situation in Germany disintegrated. The November 11 *New York Times* ran a three-column headline, NAZIS SMASH, LOOT AND BURN JEWISH SHOPS AND TEMPLES UNTIL GOEBBELS CALLS HALT. On the two-day rampage of Kristallnacht (Night of Broken Glass), more than a thousand synagogues throughout Germany were burned to the ground. The number of Jews killed in this pogrom, Europe's worst since the Middle Ages, varies from hundreds to thousands depending on whether one counts direct fatalities or ones in the ensuing year.

There was no question in either Laura's or Enrico's mind that they had made the right decision. They went into high gear now that the Nobel Prize was official. Because of his frequent travels, Fermi had a valid passport, but Laura and the children, ages seven and two, did not. Italian Jews had been ordered to surrender their passports and have race recorded in them, but Fermi, with the help of an influential friend, managed to have his wife's renewed without the damning notation. This was fortunate because Laura, already nervous, would have otherwise been far more so.

Fermi had applied to Il Duce for a needed travel permit. Asking about the delay in receiving it, Fermi wrote on December 3 to Mussolini's private secretary that it would be "a great honor to be received by Il Duce" before his imminent departure for Stockholm and then Columbia University "so I can take eventual directives on actions that I might take in scientific circles of those countries." The chances of a meeting with Mussolini on three days' notice were nil. Clearly Fermi was ingratiating himself. He did not want to impart the slightest hint to the Italian government that his exodus would be permanent.

The day before their departure on December 5, 1938, the Fermis took an extraordinary step. They went to their local parish, named after

Saint Roberto Bellarmine, the cardinal inquisitor who had summoned Galileo to Rome in 1616 and ordered him to abandon Copernican notions. The Fermis were about to undergo another kind of conversion. In the presence of the Amaldis and of Edoardo Amaldi's father, Ugo, Laura was baptized by the parish priest. Then Laura and Enrico were married in a Catholic ceremony. The official document lists Nella and Giulio as having been baptized on February 28, 1936, a week after Giulio's birth. In 1936, in the face of Germany's strict racial laws, it had been considered prudent by many to baptize Jewish children in Italy. Now in 1938, it was prudent to record a Christian marriage. In one fell swoop, the union of Laura and Enrico and the status of their children was legitimized.

On the morning of the sixth of December, Fermi attended his last faculty meeting. It was announced there that two towering figures of mathematics, Federico Enriques and Tullio Levi-Civita, would no longer be teaching at the University of Rome because they were Jewish. This was another crippling blow for Italy's contribution to that discipline.

That evening at nine, the Fermis boarded the train for the forty-eight-hour trip to Stockholm. Only Rasetti and the Amaldis were at the station to see them off. As they boarded, Rasetti said in a subdued tone, "I hope I'll see you soon." Amaldi, braced for the separation, had intuited that Fermi would have left Italy even if Laura had not been Jewish. He and Ginestra sadly bade farewell to their dear friends and respective collaborators.

Fermi had planned for every obstacle. The trip, with the exception of a short delay at the German border, went smoothly. Little Nella, sensing tension when an official slowly examined their passports, was told that everything was fine. During the two long days and nights on the train, her father's calm demeanor reassured her.

Fermi received his Nobel Prize from King Gustavus V of Sweden on December 10. In his acceptance, Fermi summarized his work of the previous four-plus years. Describing the Boys' presumed discovery of transuranics, he said they had found "one or more elements of atomic number larger than 92; we used to call the elements 93 and 94 in Rome with the names of Ausonium and Hesperium respectively." He added that Hahn and Meitner had been able to trace "elements up to atomic number 96." All those claims of transuranic elements would soon be proved wrong. Some detractors say Fermi got the Nobel for a faulty discovery. But his years of productive research, particularly showing the world the power of slow neutrons, suggest otherwise. Nevertheless, he never quite forgave himself for his mistake about transuranics.

The Nobel Prize award ceremony was simpler and shorter than usual because prizes in chemistry, physiology, and medicine were not awarded that year. Only Fermi and Pearl Buck were on the stage, the American author having received 1938's Literature Prize. At one point in the ceremonies, walking respectfully backward after accepting his prize from the king, Fermi mistakenly almost sat in Pearl Buck's lap. A quick maneuver on his part saved Buck's dignity—and his own.

The Italian press criticized Fermi for not wearing his Royal Academy uniform and for shaking the king's hand rather than giving him a Fascist salute. They relegated Fermi's receiving the Nobel Prize to their newspapers' back pages. An occasion that should been a source of national pride instead almost turned into an embarrassment. The press was pro-Hitler and didn't want to cover an award shunned by Germans. Fortunately, Fermi did not have to worry anymore about the Italian press, nor about dictates from Fascist officials.

On December 24, Christmas Eve, the Fermis sailed from Southampton to New York. The underground and the foreign antifascist press had questioned Fermi's official line that the family was leaving Italy for only six months. They reported that Fermi was leaving Italy permanently because of his wife's Judaism. A banner headline in the January 4 *New York Herald Tribune* proclaimed: FERMI, WINNER OF NOBEL PRIZE, TO SETTLE IN U.S. A subhead stated: NOTED PHYSICIST, NOW HERE, LEFT ITALY BECAUSE OF THE RECENT ANTI-SEMITIC LAWS.

The displeasure from Italy was immediate. The next day, at Il Duce's demand, Luigi Federzoni, the president of Italy's Royal Academy and a staunch Fascist, telegraphed Fermi, who had safely arrived on American soil two days earlier. Fermi's reply was as instantaneous as Federzoni's inquiry: "I thank you for the courteous communication that surprised me since it is known that my visit to America is in no way connected to racial questions but is determined only by scientific reasons, just like my preceding five visits. Please deny in my name any other interpretation." Fermi's rebuttal was forwarded to Mussolini the same day. In his transmittal memo to Mussolini, Federzoni commented that the Nobel Prize winner's reply was "without special warmth."

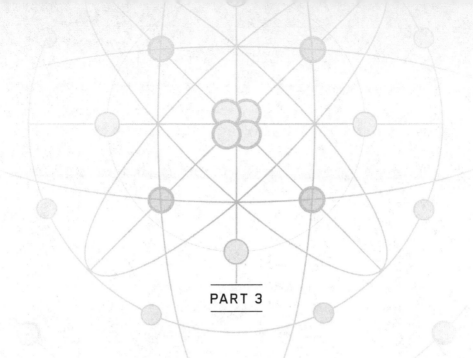

PART 3

HELLO, AMERICA

19

FISSION

The very same day, perhaps at the very same hour, that the Fermi family was boarding the ocean liner bound for America, another event took place in Sweden that would forever alter the world's scientific and human domain. It was predicated on Fermi's recognition of the power in slow neutrons.

Two physicists, an aunt and her nephew, were talking and walking in the snowy Scandinavian woods on the day of Christmas Eve. Vacationing together over the holiday, they were on the cusp of discovering the mechanism of fission. Both Lise Meitner and Otto Frisch were Jewish and part of the vast diaspora seeking refuge after the rise of Nazism. Meitner, born in Vienna, received a doctorate in physics at its university, the second woman to have ever done so. She had been in Sweden only a few months, after having escaped from Germany in July 1938. Frisch, who had also studied at the University of Vienna, was the son of Meitner's sister. He had worked first in Germany, but emigrated once Hitler was elected. For the past four years, Frisch had been conducting nuclear research at the Bohr Institute in Copenhagen.

When Meitner had moved to Berlin in her twenties, she met Otto

Hahn, a young German chemist, and the two of them combined interests in radioactivity. They would collaborate for the next thirty years at the Kaiser Wilhelm Institute for Chemistry. Meitner was charged with establishing a physics branch in the Institute. Within a decade she had been so successful that she became one of the first women to receive the title of Professor.

In the early 1930s, by now internationally known, Meitner assisted the Rome group in shifting their focus from atomic to nuclear physics. Franco Rasetti remembered her tutelage fondly. "She taught me how to prepare the polonium, how to extract and separate the polonium from radium, how to evaporate it on the beryllium foil and so on."

Apolitical like Fermi, Meitner had thought being Jewish would not harm her career and believed her conversion to Protestantism in 1908 had removed it as an issue. She was naïve: being Jewish did make a difference when Hitler came to power in 1933. The Führer's distinction was racial, not religious. According to Nazi dogma she was a Jew. Since the Kaiser Wilhelm Institute was an independent entity, Meitner's research continued unimpeded, but her teaching privileges at the University of Berlin, a state-controlled institution, were instantly rescinded.

For a few years she felt relatively safe as an Austrian rather than a German citizen. That protection ceased after the Anschluss, the March 12, 1938, annexation of her homeland. Her Austrian passport was no longer valid, and Germany, with emigration restrictions in force, would not issue her a new one. On the fourth of July, she was informed that a new policy prohibiting German scientists from leaving would soon be enforced, essentially trapping her. Her appointment at the Kaiser Wilhelm would surely be terminated soon. The vise was tightening.

Eight days later on July 12, after elaborate planning and clever stratagems, Meitner illegally crossed a remote border into Holland. She had worked at her laboratory until eight that evening, warding off any suspicions about her escape. At age fifty-nine, with only ten German marks in her pocket, she leaped into the void of uncertainty that faced all such refugees. Her only security, other than her formidable talents, was a diamond ring given to her at the last minute by Hahn; it had been his mother's.

From Holland, Meitner flew to Copenhagen and continued on to

Sweden, where Bohr had arranged a position for her in Stockholm. She was safe, but work would not be the same since the research facilities were far inferior to those she had left behind. When Laura Fermi met her in Stockholm in December, she thought Meitner was "a worried, tired woman with the tense expression all refugees had in common."

Meitner shared not only family roots and religion with her young nephew, but also—perhaps more important—a love for physics. When Frisch joined Meitner in Kungälv, a small town in southwestern Sweden, on the morning of December 23, 1938, he found his aunt rereading a letter she had received from Hahn three days earlier. She asked her nephew to read it. Meitner had absorbed its contents but was still trying to discern their implications.

In Berlin, Hahn and his younger associate Fritz Strassmann had been studying the effects of bombarding uranium with slow neutrons. The experiments had revealed an anomalous end product occasionally emerging: a radioactive isotope of barium. Since the barium nucleus has only 56 protons while uranium has 92, this seemed impossible. How could absorbing a neutron cause a uranium nucleus to change by 36 units of electrical charge? Meitner had written back to Hahn immediately, "Your radium results are very startling. A reaction with slow neutrons that leads to barium!" Hahn and Strassmann themselves realized how startling it was to find barium. When they published their finding, they admitted it went against all previous experiences in nuclear physics.

After he read the letter from Hahn, Frisch told his aunt that Hahn must have made a mistake. She replied that the experienced chemist was too careful for that. But if that wasn't the case, what did the experiment mean? On December 24, aunt and nephew went out for a walk in the quiet woods near the inn to clear their heads and discuss the incongruity. And there, they came up with a possible solution.

The conventional picture of the nucleus was that strong interactions create an effective barrier, essentially preventing anything more massive than an alpha particle from exiting. Two protons might escape but not thirty-six.

Frisch and Meitner were familiar with Bohr's recent conjectures about a new picture of the nucleus. Far from visualizing it as a compact solid sphere, he was describing it in August 1938 as "a drop of fluid, and the states of excitation can be compared with the oscillations in volume and shape of a sphere under the influence of its elasticity and surface tension." The two asked themselves, What might be further consequences of viewing the uranium nucleus as a drop? Being struck by a neutron might split the drop into two smaller ones. Were that to happen, the electric force between them would rapidly drive the two drops away from each other, gaining energy as they receded.

Aunt and nephew interrupted their walk, sitting down on the trunk of a fallen tree. They began to calculate the amount of energy the drops would accumulate in flying apart. It seemed immense, more than twenty times the energy needed to detach a neutron from a typical nucleus and more than twenty million times as much as the typical energy of a chemical bond. They conjectured this might be possible, but before believing it, they knew they had another problem to solve.

Even though Einstein's famous $E=mc^2$ asserts that mass and energy are interchangeable, the overall conservation of energy is a bedrock physics principle that cannot be violated. This principle was nothing short of gospel in the bible of physics. The picture of drops separating could only be right if there were a source for the energy of motion the drops acquired. Meitner's thirty years of experience with nuclei became relevant. She realized that the sum of the two fragments' masses was less than the uranium mass. A rapid estimate showed her that the difference in mass was roughly equal to the energy of motion needed. Everything checked out.

Meitner and Frisch understood that a uranium nucleus bombarded by slow neutrons can split into two smaller nuclei. Frisch would soon give the process a name suggested to him by a biology colleague: nuclear fission. Fermi, the other Boys, and several scientists after them had all been wrong about seeing transuranics. What the Boys had actually produced was far more important. They had failed, however, to recognize it.

Frisch and Meitner went over the details of their estimates during the next week. It repeatedly seemed correct, but they thought it would be wise to consult Bohr before either writing to Hahn or submitting anything for publication. On the first of January, Frisch went back to Copenhagen; on the third he spoke to Bohr. The Dane was preparing for a trip to the United States and didn't have much time, but not a lot was needed. As soon as Frisch began to explain what he and his aunt had concluded, Bohr slapped his forehead with his hand, exclaiming, "Oh what idiots we all have been! Oh but this is wonderful! This is just as it must be! Have you and Meitner written a paper about it?"

The third had been a Tuesday. Over the next three days, aunt and nephew, he in Copenhagen and she back in Stockholm, conferred by telephone on the contents of the paper. Frisch wrote an outline and on the evening of the sixth of January brought it to Bohr, who approved its contents. The next morning, Frisch handed Bohr a typed version of the paper's first two pages just as Bohr and his nineteen-year-old son were catching the train for Goteborg to board an ocean liner for the transatlantic crossing.

When saying goodbye to Frisch, Bohr told him that he would not speak of these new results until he had heard that the Frisch-Meitner paper was in press. He wanted to make sure they received full credit. This wasn't merely a question of protecting priority. Despite her prestige, Meitner's position in Stockholm was precarious; she had been coolly received, and she had little funding and almost no laboratory equipment. As for Frisch, he had only a temporary appointment in Copenhagen. With Europe on the brink of war, the ever-supportive Bohr was hoping to enhance their chances for more secure positions.

The two pages Frisch handed Bohr included a description of the experiment he intended to perform to establish that fission was taking place. It was relatively easy to conduct since he had the key components: a uranium sample, a neutron beam, and knowledge of what to look for. All he needed were two pieces of inexpensive and readily available equipment: an ionization counter and a linear amplifier.

Frisch had confirmation of the expected result by the end of the week. The uranium nucleus could be split. Frisch proceeded to write two letters to *Nature*, a short one on the actual experiment and a more thorough one with Meitner on the whole idea of fission. In the longer letter to *Nature*, an extension of the draft given to Bohr, Frisch and his aunt stated in no uncertain terms that if fission was correct, transuranic element detection needed to be revisited—thus bringing into question a four-year history of scientific research and its conclusions.

Frisch was in an excellent mood on the sixteenth of January as he mailed the two letters to *Nature*. He knew this was a significant piece of research, though he had no sense of how pivotal it would become. In addition, two days earlier he had received the news that his father, who had been imprisoned in Dachau after Kristallnacht, was going to be released. He felt enormous relief. All that remained was to sit down and write Bohr a report. That would wait a few days, because he was totally exhausted.

It may seem implausible that fission hadn't been detected previously since, as Frisch had just demonstrated, it was relatively simple to observe. There was, however, a justification as to why nobody had looked for fission. They were following a wise, but not always correct, scientific precept: don't go chasing something revolutionary unless there is a good reason for doing so. In 1934 the notion of a nucleus splitting in two seemed much too far-fetched to contemplate. Four years later, thanks to Bohr's picture of the nucleus and the Hahn-Strassmann result, it had a hint of logical pursuit.

Not having discovered fission embarrassed Fermi, the discomfort heightened by his trip to Stockholm that coincided with Hahn's announcement and with the Frisch-Meitner discovery. The timing was such that he even had to add an addendum to his Nobel Prize lecture, stating how the recent Hahn-Strassmann experiment made it necessary to reexamine the subject of transuranic element production.

The matter was made even more painful by the fact that four years earlier, the Boys had performed the same experiment as Frisch recently had, with one small but crucial modification. Wanting to make sure they counted only the alpha particles produced by neutron bombardment of uranium and not those due to the natural decay of the uranium sample, they had covered the uranium with a very thin foil of aluminum. It was thick enough to absorb the sample's natural decays but not so thick as to mask the expected signal from neutron absorption.

Unfortunately, the foil absorbed the massive fragments produced by fission. Had the Boys left the foil off even once, they would have seen the unmistakable pulses that Frisch observed four years later. But they may not have suspected the significance of what they were seeing without Bohr's picture of the nucleus as a drop. They might simply have decided it was an experimental error. Speculating many years later on the discovery of fission, both Amaldi and Segrè were hesitant about whether they would have recognized the meaning of the pulses. Current information, erroneous about the chemical behavior of elements near uranium on the periodic table, had helped convince the Boys they were seeing transuranics. Many physicists conjecture that if Fermi had seen a large pulse, he would have suspected something and not rested until he had found the explanation.

The oversight continued to rankle Fermi. He always prided himself on not jumping to conclusions, on preparing for all possible eventualities, and on never leaving things to chance. But in this case he felt he had erred dramatically. Later, when a University of Chicago colleague pondered the derivation of an innocuous bas-relief figure over the entrance door to the Institute for Nuclear Studies, Fermi halfheartedly joked that the figure was "probably a scientist not discovering fission."

How would the world have been different had fission been discovered in early 1935? The most frightening scenario, and not an unreasonable one, is that Hitler's Germany would have recognized fission's potential, mobilized its scientists, and embarked on a crash program to develop the atom bomb. And they might have been successful. If so, how would Europe have responded to such a threat? Perhaps Fermi's not discovering fission is one of the world's greatest gifts of good fortune.

One cannot leave the story of fission's discovery without asking why Lise Meitner did not share the Nobel Prize in Chemistry awarded to Otto Hahn in 1944 "for his discovery of the fission of heavy nuclei." Looking at the history of women scientists, there is a serious concern that Meitner was not fairly recognized. Prejudice against women may have played a part, although Madame Curie had already been honored with two Nobels. Antagonism by Manne Siegbahn, an influential Swedish physicist who also directed the institute where Meitner worked, could also have been a factor. He was very conscious of his own prestige and perhaps jealous of her recognition. In addition, Sweden may have wanted to underscore its wartime neutrality. It had awarded the physics prize to an American Jew, Isidor Isaac Rabi. By awarding—in the same year—the chemistry prize to a German unassociated with fleeing the Third Reich because of racial laws, the Nobel Committee perhaps felt it achieved balance.

After a quarter century of major contributions by Hahn, there was also a general feeling that his time had come for the award. However, the same was true of Meitner, his longtime collaborator. The most equitable resolution would probably have been the one Bohr suggested: a Nobel Prize in Chemistry to Hahn and Strassmann, and one in physics to Frisch and Meitner.

20

— • • • —

NEWS TRAVELS

On January 2, 1939, the RMS *Franconia*, the ship carrying the Fermis to America, came within sight of the low coast of Long Island. Soon afterward, despite the cold and a harsh blowing wind, almost all the passengers stood on deck to admire the Statue of Liberty, their symbolic welcome to the new country. For many of them, it would become their permanent home. But for the RMS *Franconia*, New York was a temporary port. In eight months she would be carrying war troops and in six years she would serve as the headquarters ship for Winston Churchill and the British delegation at Yalta. For now, in 1939, she had provided safe passage to Italian refugees: a Nobel Prize winner and his family.

Nella and Giulio, respectively seven and two, were bundled up as they were led down the gangplank by their parents and by the nursemaid the Fermis had brought with them. Laura had been resistant to leaving her "gracious way of life." Her attachment to Italy was to a lifestyle that included, in part, household help. The bigger issue was that Rome had always been her home, and the thought of leaving her family and the culture of the Eternal City pained her deeply. She had strong doubts that America could live up to her standards.

Enrico, in contrast, was grateful to be leaving. Science in Italy had been compromised by Fascism and he, more than Laura, realized how her Judaism made life precarious. As Laura confessed, "Enrico had often suggested that we leave . . . and each time I had raised objections." With extensive anti-Semitic laws promulgated in Italy in the fall of 1938, even Laura had known it was time to go. The offer from Columbia was a relief. Disembarking in the New World, Fermi turned with a grin to his wife and children and announced, "We have founded the American branch of the Fermi family."

George Pegram, the dean of the college as well as chairman of the Columbia physics department, greeted the Fermis warmly at the pier. His recruitment skills and organizational acumen had led this astute Southern gentleman to build one of the foremost departments in the United States. He had also established a strong research program in neutron physics. The addition of Fermi was a personal triumph for him.

Smiling as he struggled to understand Laura's somewhat limited English, Pegram ushered the Fermis to the King's Crown Hotel, located on 116th Street, the very heart of Columbia University. He made sure they were housed comfortably and provided information about nearby apartment rentals. Having spent the summer of 1936 teaching at Columbia, Fermi was acquainted with the neighborhood, but this was different. He had then come alone and only for a few months.

Exactly two weeks later, on the sixteenth of January, Enrico and Laura were back at the pier, this time to greet another physicist from Europe. The *Drottningholm* docked at the Swedish-American Line's Fifty-Seventh Street pier at one in the afternoon. Even before the ship reached its berth, the Fermis recognized Niels Bohr at the railing, scanning the reception crowd gathered on the dock. The fifty-four-year-old physicist's plan was to spend four and a half months at Princeton's Institute for Advanced Study, along with his son Erik, and then return to Copenhagen. Accompanying him on the voyage was Leon Rosenfeld, a Belgian physicist and a longtime collaborator of Bohr's.

It had been a rough crossing of the North Atlantic and Bohr had suffered from seasickness on the nine-day crossing. But he worked constantly. With Rosenfeld's assistance, Bohr had gone over and over the Frisch-Meitner arguments for fission. By the time they landed in

New York he was certain they were correct. However, he wanted this revolutionary discovery to remain secret, waiting until his friends' articles on the subject were submitted. Unfortunately, he neglected to let Rosenfeld know of his promise to Frisch not to speak of their work until then.

In addition to the Fermis, a twenty-eight-year-old Princeton physicist named John Wheeler greeted Bohr at the pier that Monday afternoon. Wheeler, who had spent a year at Bohr's Institute in Copenhagen from 1934 to 1935, had come to New York on a morning train to greet his former mentor, expecting to return to Princeton that day with Bohr and Erik. But they decided to stay in New York for a night. Wheeler took only Rosenfeld back with him.

That evening Wheeler and Rosenfeld went to the weekly informal physics gathering for Princeton graduate students and faculty. In the course of the meeting Wheeler asked Rosenfeld for news from Copenhagen. The Belgian innocently reported both the Frisch-Meitner conjecture and Bohr's impression of their work.

Arriving in Princeton a day later, Bohr was aghast to discover the physics community abuzz about the phenomenon that would soon be known as fission. He could hardly blame Rosenfeld for the leak. Still determined to protect the scoop of Frisch and Meitner, Bohr's anxieties heightened when he found no mail. Frisch had told him that he would contact him after conducting a small experiment in Copenhagen to confirm his and Meitner's findings. Had something gone amiss? Yet Bohr was confident enough about their discovery to pen a short letter to *Nature* designed to protect the precedence of the Frisch and Meitner results. On January 20, with no news from Copenhagen, Bohr wrote to Frisch inquiring about the status of things.

In a postscript, Bohr added that the Hahn-Strassmann article in *Naturwissenschaften* about the appearance of barium after neutron bombardment of uranium had just reached Princeton and was stimulating a great deal of discussion. So were Frisch and Meitner's findings on fission. On January 24, with still no word from Frisch, Bohr wrote him again. He did so with even greater urgency regarding publication of their results. An event was about to take place that would spread the news everywhere about the possibility of fission.

The event was the brainchild of George Gamow, a tall, fun-loving, and imaginative Russian, who in 1928 had been one of the first scientists to apply quantum mechanics to a nuclear physics problem. After a series of adventures, he had arrived in the United States in 1934 and accepted a faculty position at George Washington University in the nation's capital.

Having spent three years at Bohr's Copenhagen Institute, Gamow had come to appreciate the stimulation of its annual weeklong gatherings and decided he wanted to start something akin to them in the United States. Though maintaining the informality, he envisioned a different format: announce the topic ahead of time and then gather a group of thirty or so experts for three days of free-ranging exchanges. With the support of the university's administration and the Carnegie Institution as cosponsor, the first such meeting was held in April 1935. It was an enormous success and subsequently instituted on a yearly basis.

Gamow and his fellow refugee Edward Teller, who had joined him on the George Washington University physics faculty in the fall of 1935, were planning to hold the fifth such conference on January 26, 27, and 28, 1939. A topic had been chosen and an invitation list finalized. When Gamow heard that Bohr and Fermi were both on the East Coast, they were quickly added to the list. Eager to meet American physicists and other refugee scientists, many of whom were friends, the two recent arrivals accepted.

Bohr reached Washington on the evening of the twenty-fifth and had dinner with Gamow. After dropping him at his hotel, the spirited Russian immediately called Teller. His first words to his colleague were "Bohr has gone crazy. He says the uranium nucleus splits." Gamow and Teller decided to throw out the previous schedule: the conference's first two talks would instead be given by Bohr and Fermi, each explaining what it means that a nucleus can split.

That afternoon, before Bohr and Fermi went to Washington, there had been a somewhat comic wild goose chase between them. Bohr had stopped by the Columbia physics department looking for Fermi but hadn't found him in his office. Having heard the news from Princeton about fission, Fermi was talking to his colleague John Dunning about conducting an experiment like the one that, unbeknownst to Fermi, Frisch had already performed. Meanwhile, still looking for Fermi, Bohr had gone down to the department's basement, where the cyclotron was housed, hoping to find him at work there. The only person he met there was a graduate student, Herbert Anderson.

Bohr was not one to make distinctions between young and old or students and senior faculty when he felt the need to talk about physics, which seemed to be most of the time, albeit with various degrees of clarity. As Anderson later remembered, Bohr walked right up to him and started out, "'Young man, I'd like to tell you about fission.'" So I said 'Fine.' I was stupefied. Here was the greatest physicist in the world and he comes over to you . . . And Bohr, when he talks to you, he talks in a whisper. So what he does is he practically hugs you and speaks in your ear, so it's a very close encounter. And I was overwhelmed by all that." Having finished talking to Anderson, Bohr left to catch the train for Washington, never having connected with Fermi but happy to have had with Anderson what he thought was a meaningful exchange on the topic of fission.

Bohr was known for his somewhat wooly delivery. His brother Harald, a distinguished mathematician and a famously lucid lecturer, once responded to the question why his brother was such an unclear speaker: "Simply because at each place in my lecture I speak only about those things which I have explained before, but Niels usually talks about things he means to explain later."

Anderson had not understood much of what Bohr had said to him, but he perceived that it was weighty and Fermi should be informed. He rushed upstairs to the physics department to tell him what he had heard. As he later confessed, "Nothing like having a reason to talk to one of the really great guys who knew more about neutrons than anybody else in the world." He soon found Fermi. Starting to recount the conversation with Bohr, he was quickly interrupted. As Anderson remembered:

So Fermi says 'let me tell you about fission.' Then he described it, and then I really understood it. With Bohr it didn't make any sense to me at all. But when Fermi explained it, he really made it very clear. He knew, he'd already heard about it. With Fermi, all you have to know is to tell him what to think about and he knows how to take it from there and work everything out.

Fermi did more than succinctly explain to Anderson the conjecture that a uranium nucleus can be split. Fermi told him what would be needed in order to perform the key experiment. Anderson suddenly realized that he had all the necessary equipment in the physics department's basement. Summoning up his courage, he suggested they do the experiment together. Fermi, still unaware that Frisch had already conducted the experiment, instantly agreed.

This was the beginning of both a working relationship and a close personal friendship that would last until Fermi's death. As Laura Fermi wisely observed, their closeness was partially due to Anderson's boldness and lack of need for praise. Fermi saw no reason to express appreciation for a job well done. It was to be expected. This bothered some people, but not Anderson.

Fermi and Anderson's very first collaboration got off to a rough start because the basement cyclotron that was supposed to provide them with a neutron source wasn't functioning properly. Fermi had to leave for Washington before it could be fixed. Undeterred, Anderson located some radon and beryllium in the department to make a source like the one Fermi had used in 1934. That night, having prepared it and now joined by Dunning, Anderson did the experiment. The two Columbia physicists were the first in the United States to see the gigantic pulses that fission causes. They immediately sent Fermi a telegram describing their success. Excitement was building.

The next day, January 26, the Washington meeting began at two in the afternoon. Following the revised schedule, Bohr spoke first and Fermi second, his diction accented but his meaning crystal clear. Richard Roberts, an experimental physicist who worked at the Carnegie Institution, remembered that Bohr "mumbled and rambled so there was little in his talk beyond the bare facts. Fermi then took over and gave his usual elegant presentation including all the implications." With a jolt Roberts realized that he too had all the equipment necessary for doing the experiment. Roberts and a friend went right to work looking for splitters, the term Bohr was using, because he had not yet heard from Frisch the name that would be used ever after: fission.

By Saturday night, Roberts and colleagues alerted their Carnegie boss that they were ready to show anyone their results that evening. Fermi and Bohr, the latter smoking an after-dinner cigar, saw for the first time the giant pulses produced in the ionization chamber. And the whole world soon was informed about splitters because a reporter present at the Thursday afternoon talk filed a report for the *Washington Evening Star*. The Associated Press picked it up and so did the Sunday *New York Times*.

The exuberant press coverage added to Bohr's anxiety because Frisch and Meitner's role was not mentioned. What if the news item were to be picked up by the Scandinavian press? How might Frisch and Meitner feel if their scoop was not rightly credited? Still not having heard from Frisch, Bohr returned to Princeton on Sunday. There, to his great relief, he found a letter from one of his sons in Copenhagen that mentioned in an offhand way that Frisch had been successful in carrying out the experiment and had submitted letters to *Nature*.

Two days later, on January 31, Bohr received a long-awaited telegram from Frisch: "LINEAR AMPLIFIER DEMONSTRATES DENSELY IONIZING SPLIT NUCLEI BOTH URANIUM THORIUM DETAILED INFORMATION POSTED." A detailed letter from Frisch arrived the next day. Bohr wrote back at once, "I need not say how extremely delighted I am by your most important discovery, on which I congratulate you most heartily."

The personal letter to Bohr also explained why he would be using the term "fission" to describe the phenomenon: "It was suggested by the biochemist Dr. Arnold, who told me it was the usual term for the division of bacteria." Bohr and others concluded that it was better to speak of "fission," with its somewhat esoteric origin, than of "splitters," a prosaic word with electrical associations.

Although Bohr was very relieved to hear from Frisch, he still worried about whether Frisch and Meitner would be appropriately credited. He was also feeling guilty that this might have been partially due to his not having warned Rosenfeld to remain silent. Bohr tried to remedy the situation as best he could by urging that any publication on fission wait to be submitted until the publication of Frisch's letter to

Nature. These efforts were not successful, and led to tensions between Bohr and some other physicists, including Fermi.

In Fermi's first publication since coming to Columbia, a *Physical Review* letter entitled "The Fission of Uranium" written with Anderson, Dunning, and two other Columbia colleagues, the only mention of Frisch's experiment is a paragraph included after the description of what Fermi and his colleagues had done. It says, "After this experiment had been performed, Professor Bohr received a cable from Dr. Frisch stating that he had obtained the same results some days before."

As for Fermi's point of view about the discovery of fission, it was stated at the beginning of the Columbia article:

> The phenomenon was discovered by Hahn and Strassmann who were led by chemical evidence to suspect the possibility of the splitting of the uranium nucleus into two approximately equal parts. Through the kindness of Professor Bohr we were informed of these results some days before receiving them in published form, as well as the suggestion of Meitner and Frisch that the process should be connected with the release of energy of the order of 200 Mev.

In other words, Fermi was maintaining that Hahn and Strassmann had discovered fission and that Frisch and Meitner had only estimated the energy released during the process. Bohr felt this assessment was unfair. He suggested that an acknowledgment of Frisch and Meitner be more explicit in the article. Fermi replied in a March 1 letter that he would have been happy to accommodate Bohr, but it was too late to do so. The article was already in press.

Bohr's opinion, as expressed in a response letter to Fermi, tried to explain again his position, namely that Frisch and Meitner's "merit was to have grasped the fission idea so thoroughly and given so reasonable an explanation of the mechanism of energy release that it would appeal immediately to the interest of all physicists."

This difference of opinion caused temporary friction but did not leave a lasting strain between Bohr and Fermi. Each greatly respected and appreciated the other's strengths, even if they did not always see eye to eye. Bohr's approach was certainly more generous, anxious to go beyond the basic physics to the human side of the equation. Fermi's approach was based on facts directly laid before him. Fermi did not look for subtleties. Bohr was always searching for elusive truths. One of Bohr's favorite aphorisms was that a great truth is one whose opposite is also a great truth. One cannot imagine Fermi voicing such a statement.

The discovery of fission had a huge impact on nuclear physics. Under the influence of the new understanding, nuclear physics was advancing at a dizzying pace. Groups everywhere were studying fission, a phenomenon that seemed unimaginable only a few weeks earlier. The nuclear physics community looked to its two leaders, Bohr and Fermi, one for his unparalleled knowledge of theory and the other as the greatest experimentalist and clearest expositor.

Bohr and Fermi were admired and loved by the physics community. Aside from his stature as a physicist, Bohr's presence meant so much because, as written in a centenary volume commemorating him, "No great human concern left Bohr indifferent." He was a humanist and a philosopher as well as a scientist. Fermi, by comparison, was admired and loved for his single-mindedness and the purity of his devotion to physics.

21

- - -

CHAIN REACTION

The world of nuclear physics had been transformed in a few short weeks. Everybody involved in fission and a host of newcomers were speaking of it, what it meant, and how they could participate in its auspicious potential. The anticipation was palpable. A few scientists were beginning to envisage with dread the possibility that this new source of energy, dwarfing in magnitude anything previously known, could make a weapon of unparalleled destructive power.

At the end of 1939 the two acknowledged leaders in studying fission, Bohr and Fermi, also saw their respective interests in the subject move in different directions within this broader context. Bohr's focus was on anomalies in the different modes of fission and Fermi's on commonalities in all modes of fission. Fermi postulated it was likely that extra neutrons would be released when fission occurred. He began speculating about what might happen if a neutron beam were aimed at a large number of tightly packed uranium nuclei. For example, if an initial fission were to produce two neutrons, those two would generate four neutrons by colliding with other uranium nuclei. From 2 to 4 to 8 to 16 to 32 to 64 to . . . It would be a chain reaction. And if each collision generated the amount of energy Frisch and Meitner had estimated a month earlier, the end product would be a mammoth source of power.

Following this idea to its logical conclusion, Fermi realized that such a chain reaction could result in a hugely powerful bomb. The challenge was whether the energy that was produced would blow the device apart too quickly for the chain reaction to be sustained. George Uhlenbeck, his friend from Rome, Leiden, and then Michigan, found Fermi contemplating this prospect in the spring of 1939. Visiting Columbia for the semester and sharing an office with Fermi, Uhlenbeck saw him one day staring out of their large office window at the view below him. Holding his hands cupped together as if there were a small metal sphere between them, Fermi turned toward Uhlenbeck. Looking first at the imagined sphere in his hands and then down at the Manhattan landscape he quietly said, "A little bomb like that and it would all disappear." The bomb might be little in size, but its power would be earth-shattering.

Fermi put such notions out of his mind: it was too remote a possibility. For the moment his attention concentrated on the goal of producing a chain reaction, not its application. He was keen to get started. On returning from Washington, Fermi asked Anderson to join him in tackling the problem and showed him a list of experiments they would need to perform together. He had already formulated a plan of action.

Nor was Fermi the only physicist considering chain reactions in early February. On Monday morning, January 30, a young physicist was sitting in the barber's chair at the Berkeley Student Union reading the day's newspaper. Tall and blond, Luis Alvarez had grown up in San Francisco and gone to college and graduate school at the University of Chicago, demonstrating there his drive, originality, and technical wizardry. He would become a major developer of tools for both radar and the atom bomb, even flying over Hiroshima in a B-29 trailing the *Enola Gay*. Though he was only twenty-seven years old at the beginning of 1939, Alvarez had already established a reputation as a star among the experimental physicists in Ernest Lawrence's cyclotron group.

That Monday in the *San Francisco Chronicle*'s second section, Alvarez came across a small item that said "some German chemists had found that the uranium atom split into two pieces when it was bombarded with neutrons." Instructing the barber to stop cutting his hair, Alvarez pulled off the sheet clipped onto him and started running to his laboratory. On his way there he met the campus's brilliant theoretical physicist, a thirty-four-year-old New Yorker named J. Robert Oppenheimer. Stopping for a minute, Alvarez told him what he had just read and how he was on his way to produce fission. Oppenheimer's immediate response was that splitting the nucleus was impossible. Alvarez remembered Oppenheimer's giving him "a lot of theoretical reasons why fission couldn't really happen."

That did not deter Alvarez. He rapidly succeeded in setting up and performing the experiment that revealed fission of uranium nuclei. He then invited Oppenheimer to come to his laboratory and observe results for himself:

> When I invited him over to look at the oscilloscope later, when we saw the big pulses, I would say that in less than fifteen minutes Robert had decided that this was indeed a real effect and more importantly, he had decided that some neutrons would probably boil off in this reaction, and that you could make bombs and generate power, all inside of a few minutes . . . it was amazing to see how rapidly his mind worked and he came to the right conclusions.

Within a few years a much larger physics community would be commenting on the speed with which Oppenheimer seemed able to embrace all the implications of fission.

Though neither Fermi nor Oppenheimer knew it at this time, a visionary Hungarian physicist had been worrying for more than five years about the possibility of a nuclear chain reaction being used to construct a bomb. Until January 1939, he was unaware of an ingredient that made the prospect both more likely and more frightening: fission.

Forty-year-old Leo Szilard had extraordinarily acute political antennae. Jewish, he had fled Hungary as a twenty-one-year-old in 1919 to avoid persecution by the right-wing anti-Semitic government that had come to power. Moving to Berlin, he had rapidly distinguished himself through his insights in theoretical physics and his fondness for new technological instruments. He became a good friend of Albert

Young Fermi (middle)
with his siblings,
Giulio and Maria, 1905.

Letter written by Fermi from
Göttingen to his friend
Enrico Persico, March 1923.
Both drawings are meant to
be comical. One is a mock-up
of an electron hitting an atom;
the other is a caricature of a
female Göttingen physicist.

Hiking with friends, the
only three physics students
enrolled at the University of
Pisa, in the nearby Apuanian
Alps, 1921. Left to right:
Fermi, Nello Carrara,
Franco Rasetti.

Laura Capon in 1924. "[You have] no idea how beautiful the teenage Laura was," Enrico commented many years later to a female colleague.

The stars of *Knabenphysik* (boys' physics), whose brilliant research changed physics: (left to right) Fermi, Werner Heisenberg, and Wolfgang Pauli at the Volta Conference on Lake Como, 1927. Pauli, age twenty-seven, was the oldest of the three.

Fermi entering data in one of the many notebooks he wrote and then relied upon for his observations and learnings, 1928.

The wedding of Fermi to Laura Capon with friends and family on the Campidolgio steps in Rome. Note Laura's father, the admiral, in his white suit and Orso Corbino, Fermi's mentor, to the admiral's left. July 1928.

The Boys of Via Panisperna, the famous group that led the field of neutron scattering. Left to right: Oscar D'Agostino, Emilio Segrè, Edoardo Amaldi, Franco Rasetti, and Fermi, 1934.

Fermi holding his first born—Nella—in the Dolomites, summer 1931.

Fermi and American novelist Pearl Buck at the Nobel
Prize ceremony in Stockholm, 1938.

The new Americans: the Fermi family arriving in America. Giulio, age two (left),
and Nella, age seven, on January 2, 1939.

A drawing of the first atomic pile in the squash courts under Stagg Field, University of Chicago, December 2, 1942.

University of Chicago gathering on the fourth anniversary of the first pile experiment, with many of the key physicists present. Front row, left to right: Enrico Fermi, Walter Zinn, Albert Wattenberg, and Herbert Anderson. Middle row, left to right: Harold Agnew, William Sturm, Harold Lichtenberger, Leona Woods Marshall, and Leo Szilard. Back row, left to right: Norman Hilberry, Samuel Allison, Thomas Brill, Robert Nobles, Warren Nyer, and Marvin Wilkening. December 2, 1946.

The Los Alamos main gate to the secret city (around 1944). All residents of the Atomic City were required to have entry badges. The original gate was torn down in 1947 and replaced by a modern one. The city became open (no longer gated) in 1957.

A Los Alamos laboratory lecture: (left to right) Norris Bradbury, John Manley, Fermi, and Jerome Kellogg in the front row; J. Robert Oppenheimer and Richard Feynman are visible in the second row. 1944.

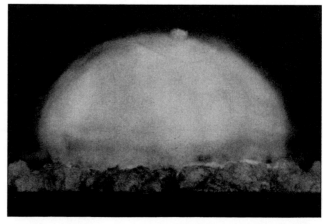

Explosion of first atomic bomb 0.053 seconds after detonation. Trinity, New Mexico, July 16, 1945.

Fermi and Emilio Segrè, his first graduate student and life-long colleague who edited Fermi's complete works. This photo was taken a week after the bomb was dropped on Nagasaki. August 15, 1945.

Fermi, whose brilliance as a teacher was much heralded by undergraduate and graduate physics students, at the University of Chicago in the 1950s.

Enrico and Laura in 1953, he with his legendary slide rule and she with a page from her forthcoming book, *Atoms in the Family.*

A family vacation in the
Dolomites with dear friends:
(from left circling to right)
Enrico Fermi, Giulio (Judd) Fermi,
Ginestra Amaldi, Laura Fermi,
Edoardo Amaldi, Ugo Amaldi.
August 1954.

The obelisk at Trinity, the remote
test site in the New Mexico desert
where the world's first nuclear bomb
was exploded on July 16, 1945.

The monument *Nuclear Energy* by
sculptor Henry Moore at the University of Chicago, dedicated to the
memory of Fermi's experiment that
introduced the first controlled generation of nuclear power.

Einstein, a man who shared both those interests with him; their partnership led to the joint filing of several patents, none of which became very profitable. Nonetheless the small amount Szilard received from his royalties subsidized the nomadic life he had chosen.

When Hitler came to power at the end of January 1933, Szilard—like Einstein—left Germany. This time he went to London, booked a small hotel room, and continued his previous lifestyle: a leisurely breakfast followed by two or three hours soaking in a tub. He would jot down on a pad ideas that had come to him while in the bath and then go about his business for the rest of the day.

In September 1933, Szilard had an insight that foreshadowed the work Fermi and his group carried out almost a year later and that contained nuggets of the problems that would fill the minds of physicists five years later. Szilard asked himself what might happen if neutrons instead of alpha particles were used as projectiles for the bombardment of nuclei. He proceeded to reflect on the prospect that this might produce extra neutrons. As Fermi and Oppenheimer had, he wondered what would be the result if this occurred amid nuclei packed together.

On the fifteenth of March 1934, coinciding with the time Fermi was writing his first paper on radioactivity induced by neutron bombardment, Szilard filed his patent for a chain reaction. It was his usual practice for concepts he found promising. Aside from his work with Einstein, he had already followed this procedure for the electron microscope and the cyclotron. Some accused him of doing this for personal gain rather than in the interest of advancing science, but that didn't bother him.

Szilard was, however, irritated that nobody seemed willing to pursue the notion of a chain reaction. He needed help and funds but could not find anybody willing to assist him. Szilard's fellow Hungarian Eugene Wigner was encouraging. He was a first-rate theoretical physicist with a mathematical bent, although his undergraduate degree from Budapest was in chemical engineering. He also had a practical side. However, Wigner, like Szilard, had no money and no laboratory. Chaim Weizmann, Zionist leader, chemist, and a friend, was intrigued but unable to raise the requisite funds.

In his memoir, Szilard mused on what might have happened had this not been the case. He concluded that it was a good thing:

I have often thought since that time that if Weizmann hadn't failed me, almost certainly Germany would have won the last war. For though I was fully aware of the implications and determined to try what I could to keep the experiment a secret, it is almost certain that in prewar England it would not have been possible to keep such a discovery a secret, and that neither in England nor in America would this development have been pushed with determination in which it would have been pushed in Germany, which in 1935 was fully determined to rearm and go to war.

As was usually the case, Szilard weighed the political implications of scientific inquiry.

In 1934, becoming more and more troubled about the rise of Nazism in Germany, Szilard took several steps to keep his design for a chain reaction from falling into the hands of physicists who might cooperate with Hitler. He assigned the patent for a chain reaction to the British Admiralty with the proviso it be kept secret. He also initiated a campaign to persuade physicists outside Germany not to publish their research on neutrons. This request, coming from a relative unknown in the field, was deemed presumptuous and ignored.

Again thinking ahead, during a 1931 visit to the United States, Szilard had filled out preliminary papers for immigration in the eventuality of relocation. These became pertinent with Germany's unopposed occupation of the demilitarized Rhineland in March 1936. England was not far enough from Germany for his comfort. On Christmas Day 1937, Szilard boarded the *Franconia*, the very same ocean liner the Fermis were to sail on a year later. And like them, he arrived in New York on January 2 and checked in to the King's Crown Hotel on 116th Street. There were, however, notable differences between the welcomes accorded to him and to Fermi. Szilard didn't have a Nobel Prize, had no obvious major accomplishments, and didn't have a position or any likelihood of obtaining one soon. He seemed curiously unfazed by any of this.

But he continued to be bothered that his chain reaction idea seemed frozen. In January 1939, having heard that Wigner, by then a professor at Princeton University, was in the school's infirmary with a case of jaundice, Szilard decided to cheer him up by paying a visit. As soon as he entered the infirmary room, Wigner told him of reports about fission circulating in the Princeton physics department. Szilard found the news electrifying. Less than a month earlier he had written a letter to the British Admiralty suggesting that his chain reaction patent be withdrawn since nothing was moving. He promptly cabled them to disregard his recent letter.

Having caught the flu in Princeton, Szilard did not return to New York City until the end of January. By then the news was out: the meetings at George Washington University, convened by Gamow, had spread the word about fission. He also heard that Fermi was set on creating a chain reaction.

With his political sensitivities aroused, Szilard rushed to meet with the Polish-born Columbia physicist Isidor Rabi. Imploring him that anything to do with a chain reaction should be kept secret, Szilard recounted his fears that Nazi Germany would decipher its implications. The two sought out Fermi to share Szilard's apprehensions. Fermi was consistent: his inner calm was unruffled. He had no qualms about publishing results since he thought the chance of obtaining a chain reaction was remote. When Rabi asked him how remote, Fermi replied ten percent. Although previously not so, Rabi was now alarmed. He retorted, "Ten percent is not a remote possibility if it means that we may die from it."

The schism between Fermi and Szilard on how to treat information regarding a chain reaction was instantly exposed. It continued along the same lines and on many fronts in coming years. As Szilard recounts in his memoir, the men were both of a conservative bent, but had opposite views on how to proceed: "Fermi thought that the conservative thing was to play down the possibility that this may happen and I thought that the conservative thing was to assume that it would happen and take all the necessary precautions."

Unable to settle their disagreement, Fermi and Szilard resolved to proceed independently for the time being and to keep each other abreast of any progress toward obtaining a chain reaction. Szilard nevertheless continued to try to convince Fermi and others in the United States not to publish any neutron research. Nothing should fall into the hands of the Third Reich.

Tensions were somewhat defused when Fermi and Szilard heard it was going to be harder to make a bomb with uranium than they had reckoned. That opinion had been reached by Bohr and articulated in answering a question posed to him by a thirty-three-year-old Czech theoretical physicist. Georges Placzek, who arrived in New York on Saturday, February 4, was on his way to a faculty position at Cornell University. He stopped in Princeton to speak with Bohr, with whom he had worked for years in Copenhagen. An expert in neutrons, Placzek was completely up-to-date. In fact, initially critical of the fission conjecture, he had been the one to suggest that Frisch look for the pulses within an ionization chamber.

Placzek was confused by a distinction between uranium and thorium that Frisch had very recently told him about. Both nuclei were similar in their response to neutron bombardment, but if a layer of paraffin was used to slow down the neutrons, only uranium underwent fission. After reflecting for a short while, Bohr was startled to realize the probable reason for the discrepancy. Ordinary uranium, like thorium, undergoes fission if the bombarding neutrons have energies greater than a million or so electron volts. But if the neutrons are slow, the only nuclei that undergo fission are those of U-235, the comparatively rare uranium isotope with 133 rather than the conventional 136 neutrons in its nucleus.

Bohr conjectured that a weapon built largely out of U-235 was likely to work and be extremely potent. However, since U-235 is only 0.7 percent of ordinary uranium ore, separating it out would pose an impossible task. Or at least, so he thought. The two uranium isotopes are chemically identical, as are any two isotopes of the same element, so no chemical procedure could distinguish and then separate them. As Bohr commented, obtaining enough U-235 for a bomb "can never be done unless you turn the United States into one huge factory."

What Bohr could not foresee was that within a few years not the whole United States, but a significant fraction of it, would be converted into "one huge factory." By the end of World War II more than a hundred thousand Americans had been employed in a secret project to build nuclear weapons.

22

THE RACE BEGINS

On the same day, March 16, 1939, that Germany invaded Czechoslovakia, Eugene Wigner went from Princeton to Columbia to meet with Fermi, Leo Szilard, and George Pegram. His mission was to urge them to deliver a warning to the U.S. government about the Nazis obtaining nuclear weapons. Trying to think of the best way to proceed, Pegram decided to call a friend who was an undersecretary of the navy. An appointment with navy staff was arranged for the next day in the office of Admiral Stanford Hooper, technical assistant to the Chief of Naval Operations.

Since Fermi was scheduled to be in Washington for other matters, the other three decided he was the optimal person to make the presentation. His expertise was impressive and just having won the Nobel Prize would be a definite asset. Fermi agreed to the task but was skeptical about its success. His premonition was deepened by overhearing how the desk officer announced him to the admiral: "There's a wop outside." Fermi's accent and his appearance, sometimes described as swarthy, had labeled him.

Rather than dramatically conveying the German threat, as Szilard or Wigner doubtlessly would have done, Fermi was his low-key self.

Without embellishment, he laid the facts before the navy officials. In retrospect, the group may have made a mistake in choosing Fermi since he tended to downplay dangers. Those present at the briefing apparently concluded that there was no cause for alarm. The meeting was treated as a courtesy call; no further action would be required.

Neither did Fermi see the need to act further. He was interested in the challenge of producing a chain reaction and not eager to deal with the extra complications of political or military involvement. Fermi only wished to proceed with experimenting. He was also happy to see that the experiments he and Anderson were planning to conduct were similar to the slow neutron ones he and the Boys had performed a few years earlier in Via Panisperna. Fermi was optimistic about their outcome; no extra retooling would be needed.

In their initial effort, Fermi and Anderson suspended a small spherical bulb in the middle of a large water-filled tub. The bulb contained a source of neutrons generated, as in Rome, by a mixture of radon and beryllium. The water was used to slow down the neutrons. Strips of a dependable neutron detector were placed at various distances from the bulb to see if the neutron count went up after uranium oxide was introduced. They soon detected a rise, exactly what was expected. Uranium's presence in the water had led to the creation of extra neutrons.

A possible caveat to their results called for redoing the experiment with a lower-energy neutron source such as radium. There was no Divina Provvidenza at Columbia to go to for radium, as there had been in Rome, but Szilard saved the day. The enterprising Hungarian excelled at cultivating rich individuals who might finance his schemes. One such Maecenas lent him the funds needed to rent a gram of radium for three months.

With mutual dependency, Fermi and Szilard—an unlikely pair—joined forces. The radium Szilard rented and some uranium oxide he managed to borrow allowed him and Fermi to conduct a larger experiment. Theirs was not an ideal match. Fermi insisted on carrying out his share or more of the physical labor. Szilard disdained such efforts, hiring somebody to perform those tasks. Given their dissimilarities, it was predictable that the two directly collaborated only once.

The results of the experiment spurred Fermi and Szilard on to an expanded version of their present setup. The hopes of producing a chain reaction were curbed by George Placzek, who had led Bohr to conclude that only the rare U-235 underwent fission by slow neutrons. Placzek had also spent a year doing research in Via Panisperna during the early 1930s and was eager to see Fermi again. Taking the train from Cornell, he described his current research to Fermi, who—in turn—showed him plans for the larger experiment he and Szilard were thinking of. Placzek told them it would not work. Water, with its hydrogen content (H_2O), was necessary to slow neutrons down, but it also absorbed them. They couldn't afford to have this happen if they wanted to obtain a chain reaction. Fermi and Szilard were at an impasse.

The lighter the element, the more effectively its nucleus slows neutrons. Since using hydrogen had proved unworkable, Fermi and Szilard considered the next elements on the periodic table. Helium was impractical and beryllium was dismissed as dangerous to use in large quantities. Lithium and boron were known to be strong neutron absorbers, so they were not candidates. Carbon came next. There were unknowns about how it interacted with neutrons but it might work, especially in the form of graphite, the smooth black substance running through a lead pencil.

It was going to be expensive to purchase the required amount of graphite, perhaps tens of thousands of dollars. Fermi and Szilard thought it was worth the effort but lacked the funds to do so. This time Szilard did not have a rich patron. Regrettably, their experimental program came to a standstill at the beginning of the summer of 1939.

Research on obtaining a chain reaction had come to a halt in the United States amid lingering fears that Germany was making strides toward producing an atomic bomb. There was no verification of German activity, one way or another, but the prospect was chilling. In France, Frédéric Joliot and his associates had independently performed the same fission experiment as Fermi and Szilard and submitted a note on their results to *Nature*. The race was public and the stakes were high.

Szilard was having no success in convincing American, French, and English scientists to withhold publication of fission-related research. Bohr, like Fermi, was insisting that secrecy should never be introduced in physics. The passivity of the United States was what concerned

Szilard the most. He was stupefied that "between the end of June 1939 and the spring of 1940, not a single experiment was under way in the United States which was aimed at exploring the possibilities of a chain reaction in natural uranium."

Meanwhile, during that summer of 1939, Fermi—accompanied by Laura, Nella, and Giulio—left New York to teach at the University of Michigan Summer School. Since his experimental program was stalled, he had decided to go back to being a theoretical physicist. This was neither the first nor the last time in his career he would make the switch.

Edoardo Amaldi was in Michigan as well, officially having come to the United States to study American facilities with the intent of building a cyclotron in Rome. Unofficially he had been looking for a position in America. The move would prove impossible even if he did find a position; his wife's request for a passport had been denied. Italy was now limiting exits. A dispirited Amaldi went back to Italy in early October 1939 on an almost empty steamship, knowing that he would be the only one of the Boys still there. He was feeling more anxious about the future than ever before.

Franco Rasetti had happened to sail on the same boat to America as Amaldi, but his destination was Canada. Disgusted by Mussolini, he managed in 1939 to secure a professorship at Laval University in Quebec and departed from Naples with his mother on July 2. The voyage gave the two physicists a chance to talk. Unlike Amaldi, Rasetti decided he was no longer interested in fission. When asked in a 1982 interview why he had left the field and turned to cosmic rays, his reply was simple: "Because cosmic rays are free and everywhere." Rasetti could pursue his quest in Canada, whereas Amaldi had to return to Italy's oppressive atmosphere.

Obtaining a position in the United States would not have been a problem for Werner Heisenberg, also a lecturer that summer in Michigan—both Chicago and Columbia were wooing him—but he was adamant about wanting to remain in Germany. In small and informal gatherings, Enrico, Laura, and many others pressed him, emphasizing the Nazi regime's ruthlessness and the anti-Semitism that had driven away so many of his colleagues. Heisenberg would not budge from his conviction that he had to go back to Germany.

In Heisenberg's memoir, he recounts an argument Fermi made to him for staying in America. It is particularly interesting for what it says about why Fermi liked the United States so much. What Fermi apparently told Heisenberg was "This is a big and free country where you can live without being crushed by the weight of history. In Italy I was a great man: here I have gone back to being a young physicist, which is infinitely more exciting. Throw away the ballast of the past and begin again."

In spite of his strong Italian roots, Fermi had never been inextricably wedded to his native country. As early as 1922, Fermi had stated that he was open to emigrating. When his disbelieving sister had asked, "Where to?," Fermi shrugged his shoulders and answered, "Somewhere . . . the world is large." Heisenberg was another matter. He had been a fervent adherent of the German youth movement, and his nationalism was deeply implanted. He was not seriously tempted to stay in the United States. His country needed him. His decision created scars and an abiding divide in the international scientific community.

Less than a month after Heisenberg's return to Germany in August 1939, World War II began. Whether this complicated man actually wanted his country to win World War II or intentionally dragged his feet on the Third Reich war effort are questions that have never been resolved. Heisenberg was ambivalent about the atomic bomb, as were many scientists, including some who were working on it, but the tone he adopted in writing about the Allies' use of it seemed disingenuous if not offensive. Heisenberg wrote that he "found it psychologically implausible that scientists whom I knew so well should have thrown their full weight behind such a project."

Parts of Heisenberg's book *Physics and Beyond* border on a rewriting of history. According to Heisenberg, while in Michigan, Fermi had asked him, "Don't you think it possible that Hitler may win the war?" It is dubious that Heisenberg replied—as reported in his book—that it was not possible for Hitler to triumph, adding that "Hitler is irrational and simply shuts his eyes to anything he does not want to see." Heisenberg was surely under surveillance, and his comments would have exposed him to tremendous peril.

One outcome is clear, though perhaps it would have been so even if Heisenberg had chosen another path. His contribution to physics after World War II was minimal. By contrast, Fermi's influence and contributions as both a researcher and teacher were immense. He thrived by "being a young physicist."

While Fermi focused on theoretical physics in Michigan, Szilard was redoubling his political activities. Along with Edward Teller and Eugene Wigner, Szilard sought ways of lessening the danger of Germany's developing a nuclear weapon. All three Hungarians agreed that limiting Germany's access to world supplies of uranium would be a strategic first step. There was little that could be done about the Czechoslovakian mines, already in German hands, but perhaps Belgium could be dissuaded from selling Hitler anything from its extensive uranium deposits in the Congo. Szilard remembered that Einstein was a close friend of Belgium's queen; he had taken refuge there between leaving Germany and settling in Princeton. If Einstein asked her, maybe a ban on uranium sales would ensue.

Wigner and Szilard went on the sixteenth of July to inform Einstein about the recent physics findings and to ask if he would compose a letter to the queen. Learning of the possibility of a chain reaction, Einstein famously said, *"Daran habe ich gar nicht gedacht"* ("I hadn't thought of that at all"). However, Einstein felt it would be more appropriate to go through diplomatic channels than to write directly to the queen.

That scheme changed almost instantly as a result of a meeting Szilard had with one of his influential friends. Alexander Sachs had worked on Franklin Roosevelt's 1932 presidential campaign and had been a member of the National Recovery Administration before joining the Lehman Brothers investment firm in 1936. When Szilard described the situation to him, Sachs suggested that President Roosevelt be informed posthaste. If they obtained a letter from Einstein, Sachs would make sure it reached the president.

Szilard went back to Einstein, accompanied by Teller. With Einstein agreeing on the new plan, Szilard drafted a letter for him to be transmitted to Sachs. Dated August 2, 1939, the letter warns:

In the course of the last four months it has been made probable— through the work of Joliot in France as well as Fermi and Szilard in

America—that it may become possible to set up a nuclear chain reaction in a large mass of uranium, by which vast amounts of power and large quantities of new radium-like elements would be generated. Now it appears almost certain that this could be achieved in the immediate future. This new phenomenon would also lead to the construction of bombs and it is conceivable—though much less certain—that extremely powerful bombs of a new type may thus be constructed.

Drafts of the letter had been written by Szilard and then rewritten by Einstein in German, his native tongue. The revisions went back and forth. When Szilard dictated the final version of Einstein's letter into English, an innocent secretary at Columbia concluded Szilard was "a nut." A letter from a Hungarian pretending to be Albert Einstein, writing to the president and talking about bombs? It stretched credibility and sanity.

The letter called for quick action. Its main recommendations were to secure uranium deposits, provide funding for research related to the use of uranium, and maintain a liaison between the physicists involved in this research and the government. The letter was delivered to Sachs on the fifteenth of August, but it was not until the eleventh of October that Sachs got to see the president. Fortunately Roosevelt's reaction was not the same as the incredulous secretary's. The president retorted, "Alex, what you are after is to see that the Nazis don't blow us up." Sachs replied, "Precisely."

Roosevelt rapidly authorized the formation of the Advisory Committee on Uranium, to be headed by Lyman Briggs, the director of the National Bureau of Standards. Its members included an aide of Briggs's and an army ordnance expert, Merle Tuve from the Carnegie Institution, Sachs, and—at his recommendation—Fermi, plus the three Hungarians.

The newly constituted committee met in Washington on the twenty-first of October. Convinced little would result from the meeting, Fermi decided he would not attend. He asked Teller to represent his point of view. Tuve, who had a conflict, asked to be represented by Richard Roberts, the physicist who had shown Bohr and Fermi the fission-produced pulses at the Washington Conference.

The small meeting began contentiously. Roberts claimed that it was unlikely that a chain reaction could be obtained; his opinion was that neutron absorption effects would simply be too large. The army ordnance officer was openly contemptuous of the ability of physicists to develop new weapons. He insisted it was the morale of the troops and not weapons that won a war. At that point, as Szilard remembered, the mannerly Wigner could no longer restrain himself and interrupted by saying that "it was very interesting to hear this. He [Wigner] had always thought that weapons were very important and that this is what costs money, and this is why the Army needs such a large appropriation. But he was very interested to hear he was wrong . . . And if this is correct, perhaps one should take a second look at the budget of the Army, and maybe that budget could be cut." The army officer rapidly backtracked.

In the end Sachs and the three Hungarians prevailed, but only to a minimal extent. The committee recommended purchasing graphite and uranium; the proposed amount was a very modest, almost laughable, six thousand dollars. Even that was not immediately approved. Though the war was unfolding across the Atlantic, the United States felt no urgency about rearming. There continued to be almost no incentive to develop a nuclear weapon.

The German situation was quite different. Under the directorship of the physicist Kurt Diebner, an adviser to the German Army Ordnance Office, a series of meetings in Berlin were held in September 1939 to discuss how nuclear fission might contribute to the war effort. The Berlin-Dahlem Kaiser Wilhelm Physics Institute where Lise Meitner had once worked was placed under Diebner's command.

A group studying fission and its ramifications, informally known as the Uranverein (Uranium Club), was organized and Heisenberg, who had just returned from the United States, was put in charge of its theoretical division. The Third Reich's leading nuclear physicists were informed of the latest available findings and told their research would be funded if relevant to advancing the Uranverein's mission. The German drive to employ fission was scarily already under way. Scientists were not hesitant in pushing for this agenda even though several, including Heisenberg, had experienced difficulties with the Nazi regime.

23

NEW AMERICANS

On September 1, 1939, Germany invaded Poland; two days later, France and Britain declared war on Germany. The Fermis had dreaded the outbreak of war; the suddenness with which it began stunned them. In May of that year, the Italian and German foreign ministers, Galeazzo Ciano and Joachim Ribbentrop, had signed the Pact of Steel, in which the two nations pledged to support each other in case of war. It seemed a foregone conclusion that Italy would enter the conflict on Hitler's side, but Mussolini seemed hesitant. Many still harbored a hope that Italy, like Spain, would remain on the sidelines.

Amid this uncertainty, Fermi was focused on becoming a typical American. As with almost every other task in life he undertook, Fermi worked incredibly hard and diligently at this one. As Segrè wrote of him, "Among adult immigrants, I have never seen a comparably earnest effort toward Americanization." Fermi's "earnest effort" meant attempting assiduously to get rid of his Italian accent, making frequent use of colloquialisms in his day-to-day speech, and reading regularly what he thought were the most typical American publications: *Reader's Digest* and comic strips. Fermi drew the line at following baseball, but he had not been much of a soccer fan in Italy either.

Laura's Americanization was more gradual. Her acquisition of English began largely with the age-old rituals of shopping and cooking, done in tandem with a nursemaid who had become a general housekeeper. Other than that she learned from Nella, who caught on easily to English while attending a progressive school. Giulio was more intent on having others speak like him than on his speaking like them. But both children were catching on to the meaning of America as the Land of the Free. Nella was always asking for "more freedom," and Giulio would proclaim, "You can't make me wash my hands. This is a free country."

The family's rented apartment, even though conveniently near the Columbia campus, did not quite fit the Fermis' image of middle-class society in their new world. They wanted to live the American dream. Laura recalls Enrico feeling this goal particularly strongly. To be true Americans, the Fermis needed a suburban house with a garden and good public schools for their children. Two cars would round it out nicely.

Harold Urey, a Columbia colleague and fellow Nobel Prize winner, urged them to meet all those conditions by moving to where he lived: Leonia, a New Jersey community directly across the George Washington Bridge from New York City. The Ureys were persuasive. By September 1939, back from Michigan, the Fermis were settled into a comfortable house "on the Palisades, with a large lawn, a small pond and a lot of dampness in the basement." There were flowerbeds and a rock garden. Urey encouraged Enrico to become even more American by gardening and by mowing his lawn, but as Laura said of her husband, "his peasant blood was not aroused." His assimilation had limits. Fermi preferred going for a walk or playing tennis in his free time rather than pulling up crabgrass. His utilitarian streak reasoned that it was green and covered the lawn just as well as any other grass.

By this time, it was well past the Fermis' six-month temporary leave granted from Italy. The American press had it right after all: Fermi was settling in America. No longer pretending otherwise, Laura and Enrico retrieved their furniture, still in Rome. The continuing presence of the furniture in their apartment had been noted in police surveillance reports, providing a bit of assurance that the family intended to come back to Italy. However, the Fermis had not escaped the suspicions of the police, who reported in March 1939: "We have noticed that Professor Fermi has always shown a deferential attitude toward the regime but not a great deal of enthusiasm. We don't know if and when he'll return." There was finality to their move to America when the Fermis were reunited with their furniture in Leonia. Things had been packed so hastily that a bag full of garbage was included.

Despite their efforts to fit in, within little more than a year the Fermis and the other recent Italian immigrants would be classified by the American government as enemy aliens. Italy had entered the war. When Mussolini witnessed the German invasion of Belgium, Holland, and France, he feared that postponing his decision any longer would mean missing the spoils of victory. Nothing seemed to stop Hitler's aggressions. In early June, French and English troops had been evacuated at Dunkirk in the face of German conquest. On June 10, with the Germans rushing toward Paris, Mussolini announced that Italy would fight at their side. As was often the case, both Il Duce's timing and his judgment were terrible.

Italy's entry into the war on Hitler's side added greatly to the Fermis' worries about friends and relatives, especially Jewish ones. Laura's father was a widower, her mother having died in 1935. Partially paralyzed by a stroke, he showed no interest in leaving Italy. Despite its passage of racial laws, the retired Jewish admiral remained a strong supporter of the monarchy he had served for so many years. He also believed his own safety was assured.

Fermi had not taken a public stand against the Mussolini regime, but the Italian embassy had followed his activities with suspicion ever since his departure in December 1938. With a war on, Rome wanted an indication of where Italy's most prominent scientist stood. In September 1940, the embassy sent a report on him to the Ministry of Foreign Affairs, "Prof. Fermi, though not actively engaging in an antifascist campaign, must be considered as having passed to the camp of the opponents. He does not frequent any Italian milieus and all his relations are with the intellectuals who are the most persistent enemies of our Regime." The embassy, unenlightened about Fermi's research on nuclear fission, had no idea how advantageous it would be for America that Fermi had, as they put it, "passed to the camp of the opponents."

Although Fermi's experiments on fission had ground to a halt in the summer of 1939 for lack of funds, they resumed in the early spring of 1940 when the promised six thousand dollars from the Advisory Committee on Uranium finally reached Columbia. Fermi and Szilard agreed that Szilard should lead the search for needed supplies but that Fermi would be responsible for conducting experiments. This was an understanding that suited both of them. Neither knew what the research might lead to. It seemed unlikely that it would aid weapon development, but Szilard was keeping a sharp eye on such a prospect.

The experiments' official kickoff was marked by the arrival of one and a half tons of graphite blocks. Fermi and Anderson worked side by side, assembling them into an eight-foot-high column with a nine-square-foot base. Slots were cut into fifteen of the bricks strategically located at different heights, and foils that registered neutron counts were placed in the slots. The neutron source was put at the structure's bottom and operated for one minute. The foils were then quickly removed and carried down the hall to Fermi's office, where Geiger counters measured their induced radioactivity. It was reminiscent of the process at Via Panisperna: rushing scientists, repeatedly careening down a hall. This was the kind of experiment Fermi enjoyed. He was in charge, understood all the details, and was involved at every step. Things ran like clockwork.

The purity of the graphite Anderson and Fermi used turned out to be crucial. Getting what they wanted required circumventing the Bureau of Standards, the intermediary between Columbia and the graphite's provider, the National Carbon Company. Szilard remembered a lunch he and Fermi had with two representatives of the company in February 1941 during which an embarrassed silence followed his asking about its purity, "You wouldn't put boron in your graphite, or would you?" Since the probability of boron absorbing a thermal neutron is more than a hundred thousand times as great as that of graphite, even a minute amount of this contaminant made a huge difference. The flaw was corrected.

A policy of silence on fission research had fortunately been instituted within the physics community. The editors of the *Physical Review* asked authors submitting papers related to uranium research to withhold them from publication. Szilard's plea for secrecy had finally been heard. The secrecy was judicious and timely because Walter Bothe, a distinguished German experimentalist, was in the process of measuring the absorption in graphite of thermal neutrons. The result he obtained, more than twice Anderson and Fermi's value, led him to conclude that graphite was unsuitable for slowing neutrons in uranium experiments. Had Bothe known that impurities were almost certainly present, he would have continued his explorations. Because Bothe didn't, Germany turned its attention to using heavy water for slowing down neutrons. Its progress was seriously hampered because heavy water—water in which ordinary hydrogen is replaced by its isotope deuterium—is far more arduous to obtain.

Meanwhile, Fermi and Anderson, cautiously optimistic that graphite would be effective, introduced uranium into their cagelike structure. All the experiments the two of them conducted over the next eighteen months were aimed toward obtaining a self-sustaining chain reaction induced by fission. The technical problems were innumerable, but Fermi proceeded to solve them one by one.

This was where Fermi's extraordinary ability to bridge experimental and theoretical physics came to bear. To describe in accessible terms a self-sustaining chain reaction, Fermi introduced a coefficient labeled k, his famous neutron reproduction factor. The probabilities of a neutron being slowed down, being absorbed by either uranium or by carbon and eventually producing fission, were all taken into account and multiplied together to reach a convenient single term he called k. It explicated what happened in each generation of neutron production. If k was greater than one, the number of neutrons would keep increasing, while if it was less than one, that number would decrease. A self-sustaining chain reaction would only be possible if k was greater than one.

There was one more consideration. As Fermi wrote in a postwar article, having k greater than one meant the desired reaction will "always take place provided the leakage loss of neutrons is sufficiently small. This, of course, can always be achieved if the size of the pile is large enough."

The term "pile" was coined by Fermi because the assemblage of carbon and uranium could in principle take any shape: it might be a cube, a sphere, a column, or whatever was most opportune. It was just a pile. The bigger the pile, the smaller the leakage. Fermi and Anderson realized that a self-sustaining chain reaction would require their building a much bigger pile. They would need more graphite, more uranium, and a bigger space to work in. Obtaining all of that was going to be expensive. They had no idea where the money would come from.

The government seemed to be the only possible source. The Uranium Committee founded in late 1939 had proved to be powerless and its leader, Lyman Briggs, ineffective. However, in mid-1940, the time when Fermi was appealing for more funds, a far more authoritative organization had been created within the government. The new superorganization was the National Defense Research Council (NDRC). Its creation was an American acknowledgment that scientific research would play a major role in the war effort.

The NDRC was the brainchild of Vannevar Bush, a talented engineer who left the vice presidency of MIT in 1938 to become president of Washington's Carnegie Institution. His motivation was precipitated by the desire to be closer to the seat of power and government decision making. With war in Europe and the United States' imminent entry into it, he surmised the country needed a dedicated government agency led by scientists capable of discerning what should be supported and what funding levels were appropriate. The organization was to be independent from the military and report directly to the executive branch of government.

Approaching key contacts with both scientific and administrative credentials to help found his brainchild, Bush selected three with particular care. They were Karl Compton, the president of MIT and the older brother of the Nobel Prize–winning Chicago physicist Arthur Compton; James Conant, a distinguished chemist who was president of Harvard; and finally, the physicist Frank Jewett, president of both Bell Telephone Laboratories and the National Academy of Sciences. All three accepted the invitation to join him.

It was a formidable assembly. After lobbying the army, the navy, and Congress, Bush proposed the idea to the president on June 12, 1940. The meeting was brief: the NDRC was approved and Bush was appointed its director. The group was charged with coordinating resources and advising on scientific research related to warfare. Nuclear physics fell under its purview, and accordingly Briggs's committee was subsumed as a subcommittee.

Inexplicably, Briggs did not alert the NDRC to the warnings he had been receiving from the likes of Szilard and Wigner. The upshot was that the NDRC was not even thinking about nuclear weapons. These simply were not on its radar. Conant, Bush's right-hand man and the chair of the NDRC's chemistry and explosives division, admits to not having heard of the possibility of using fission in a bomb until March 1941. He only learned of it then because Churchill's scientific adviser happened to tell him during a visit Conant made to Britain. So much for military preparedness.

The situation changed within a short time, but developing a self-sustaining chain reaction was not an NDRC priority. On the other hand, a large pile might be a useful and novel way to generate power and Fermi's stellar research was widely respected. Columbia's application for a large-scale graphite-uranium experiment was funded by the NDRC but granted only $40,000 of the $140,000 requested. The grant wasn't as large as Fermi had wanted, but it was enough for him to get started on a bigger pile.

Fermi worked under primitive conditions. He had explored the university in 1940 together with Pegram in search of the bigger space the new pile would need. The search led them "to dark corridors and under heating pipes and so on to visit possible sites for this experiment." They found what they were looking for in the basement of Columbia's Schermerhorn Hall: a large, gloomy, cavernous room with a very strong floor.

Pegram, the physics department chairman, suggested he hire some undergraduate football team members to carry fifty tons of graphite into the space for building a bigger pile. Anderson, Fermi, and a few other physicists who joined the bulky athletes began building a new column. The old column had been eight feet high and three by three feet at the base. This one was eleven feet high and eight by eight feet at the base.

Naturally, Fermi again did not flinch from taxing manual work, pushing blocks of graphite through a bench saw and disappearing into a cloud of black dust. The eight tons of uranium oxide presented other problems. It arrived as powder, the only form available for purchase. To remove impurities, it needed to be heated to several hundred degrees and, while still hot, placed in cubical metal boxes. There were in all 288 of these boxes. They were soldered shut to ensure no moisture entered and arranged in a lattice structure inside the graphite. Slots were cut for foils to measure the neutron flux throughout the pile.

Later, Fermi reminisced in a semi-humorous fashion about how, given that both graphite and uranium oxide are black, the Columbia University physicists "started looking like coal miners and the wives to whom these physicists came back tired at night were wondering what was happening." The toll collector on the George Washington Bridge may also have been curious as to why a gentleman headed for well-to-do suburban Leonia was covered with black dust.

At the end of September 1941, Fermi, Anderson, and the team had their results. The experiment worked but the result was disappointing: k was only equal to 0.87, clearly less than the k greater than one needed for a chain reaction. But it was close enough for Fermi to believe he might be able to reach his goal if he had more graphite and more uranium, purer materials, and a still larger pile. This was going to be a bigger job than his team could handle and would require a substantial input of money.

In 1941 it probably would have been impossible to obtain the funds for a much bigger pile had it not been for an intervening set of nuclear physics discoveries made in Berkeley that past year. These indicated there might be another way to make a nuclear weapon, one more promising than separating U-235 from U-238. It involved creating a new substance, a transuranic with 94 protons in its nucleus, two more than uranium.

The only way to obtain enough of it for a bomb, however, was to produce it in a Fermi-like pile undergoing a chain reaction. Far from being secondary, such a pile might be key to the war effort. Fermi would no longer pursue science for science's sake. Obtaining a self-sustaining chain reaction inextricably tied him to the evolution of nuclear weaponry.

. . .

24

THE SLEEPING GIANT

A merica was slow to awaken to the perils of the maelstrom engulf-ing Europe. U.S. isolationists, intent on avoiding enmeshment in battles fought on foreign soil, stayed firm in their convictions. The United States behaved like a sleeping giant, devoid of nightmares of being attacked and oblivious to the prospect of a mega-bomb.

Because of NDRC's small grant to Columbia, NDRC head Vanne-var Bush was aware of the progress Fermi was making. But before com-mitting additional funds to fission research, Bush sought advice from the National Academy of Sciences (NAS). He asked them to appoint a committee of experts to evaluate the field, with particular emphasis on possible applications for national defense. Arthur Compton was chosen to head the committee. He was an excellent choice even though nuclear physics was not his specialty. Universally liked and respected, he was a first-rate scientist and administrator. His 1927 Nobel Prize in Phys-ics had been for work confirming features of quantum physics, but Compton was well versed in almost all of physics.

His committee's report, the first of three he prepared during 1941, was delivered to the NDRC on the seventeenth of May. It was sanguine about using nuclear fission as a significant way to generate power and concluded that fission might have military applications. But given the difficulty of separating U-235 from U-238, the report estimated it would take several years to make an atom bomb. A second report, delivered on July 11, said much the same. The consequence was that in the summer of 1941, the NDRC was heading for a policy of passive observation. As Arthur Compton wrote in his memoirs, "The government's responsible representatives were thus very close to dropping fission studies from the war program." Fission was not high on their agenda.

That attitude began to change after committee members read a letter from Ernest Lawrence, the winner of the 1939 Nobel Prize in Physics for his development of the cyclotron. He was a committee member but had been unable to attend their July meeting. Dated July 11, the letter expressed Lawrence's concern that the committee's outlook was overly conservative. He wanted to let them know of a recent finding that signaled a promising way of building a bomb.

Almost two years earlier, a young Berkeley associate of Lawrence's, Edwin McMillan, had tried producing transuranics. Earlier research by the Boys and others was wrong, but McMillan had found a clever way to distinguish transuranics from fission fragments.

McMillan knew that ordinary uranium, U-238, turns into radioactive U-239 after absorbing a neutron. He studied two of U-239's decay modes that seemed particularly interesting. One had a half-life of twenty-three minutes, and the other, much slower, decayed in a little over two days. Both were examples of beta decay, in which the original nucleus emits an electron. More work was needed, but McMillan made a positive identification during the spring of 1940. The twenty-three-minute decay's endpoint was a nucleus with 93 protons, a genuine transuranic. Since Neptune follows Uranus in the solar system, he named it neptunium.

However, the longer decay mode was still a mystery. In the fall of 1940, McMillan, scheduled to leave California to work on radar research at MIT, turned the problem over to the twenty-eight-year-old Berkeley chemist Glenn Seaborg.

Over the next nine months, Seaborg and a new team made a series of revealing discoveries about that long decay mode, confirming the suspicion that it was because the transuranic had decayed into a second transuranic, one whose nucleus had 94 protons. Pluto followed Neptune, so this one was given the name plutonium.

Nuclear physics considerations suggested that, when bombarded by slow neutrons, a nucleus with 94 protons would undergo fission in the same way U-235 did. This raised an intriguing possibility: a second route to a super-bomb that did not call for the difficult practice of separating two chemically identical isotopes.

During a long discussion with Emilio Segrè in December 1940, Fermi confirmed the correctness of this supposition and suggested they meet with Lawrence, who happened to be in New York, and with Pegram, in order to decide how to move ahead. They agreed that Berkeley's laboratory should pursue seeing if element 94 underwent fission and if it was stable enough to be used in a bomb. Segrè was excited by the possibilities the project offered. In the next six months, his work with Seaborg's team established that plutonium, with a half-life of twenty-five thousand years, was nearly stable, and that, like U-235, it underwent fission after capturing a slow neutron.

This was the basis for Lawrence's pressing July 1941 letter to the National Academy Committee: a second route to a bomb was within sight. Prophetically, its name was associated with Pluto, the Roman god of the underworld. The myth converted, in years to come, into a monstrous reality.

Berkeley researchers had collected only microscopic samples of plutonium. The way to obtain more was through a self-sustaining chain reaction; that could be achieved if Fermi was able to build a critical pile. Realizing how pivotal this was, Arthur Compton became an energetic advocate for Fermi to achieve his goal.

This was not the only information galvanizing American support for nuclear fission research during the summer of 1941. Marcus Oliphant, a prominent Australian nuclear physicist transplanted to England, was the forceful messenger of the additional news. Speaking to high-level American scientists, Oliphant told them that British research showed the route to a U-235 bomb was far less arduous than the NDRC had imagined. Given the succession of Axis victories, Oliphant felt that the very life of Britain might depend on having this weapon. He repeatedly made the point that the United States needed to collaborate with Britain to build it. The States, soon to be similarly threatened, could not afford the false sense of security they were enjoying.

Oliphant was privy to the current status of British fission research because he was one of six elite physicists who made up the MAUD Committee, a government-appointed group formed in early 1940. Oliphant had instigated the committee's creation because two German physicists he had recruited to his Birmingham department had given him alarming news. The top-level MAUD Committee had adjudicated its validity.

The Germans were Rudolf Peierls and Otto Frisch, both Jewish refugees from Nazi Germany. In early 1940 they concluded it would not be possible to make a bomb out of U-235 using slow neutrons as a trigger; the device would heat up so much that it would come apart before fully detonating. Subsequently, in estimating the amount of U-235 that would be needed if fast neutrons were employed instead, they had come up with a frighteningly small number.

As Peierls wrote in his memoir, "We estimated the critical size to be about a pound, whereas speculations concerned with natural uranium had tended to come out with tons. We were quite staggered by these results: an atomic bomb was possible, after all, at least in principle!" Startled, Frisch and Peierls realized that a bomb of this sort might not cost much more to build than a battleship. If so, there was an imminent threat from Germany.

By the end of the spring of 1941, having studied Peierls and Frisch's analysis, MAUD was prepared to file its dire conclusions. A complete draft was forwarded to Lyman Briggs and to Vannevar Bush on July 15. The recommendations were crisp and unequivocal: (1) developing a uranium bomb was feasible; (2) such a bomb was likely to be decisive in a war; and (3) building it should be given the highest priority. The committee also provided technical details about the amount of U-235 needed for a bomb, estimates of the cost for separating it from uranium ore, and the expected destructive power it would have. MAUD vigorously endorsed continued collaboration with the United States.

In August 1941, Oliphant flew to the United States. There were official reasons for his trip, but uppermost in his mind was his wanting to know why there had been no response to the MAUD Committee's draft from either Briggs or Bush, both of whom had seen it. The answer was that apparently neither had heard the report's clarion call.

Bush was waiting for confirmation of its contents. Briggs, whom Oliphant later described as "inarticulate and unimpressive," had simply locked the report in his safe without showing it to his NDRC subcommittee on uranium. A government employee for more than forty years and now head of the National Bureau of Standards, Briggs had treated the MAUD report as nothing more than one of the many government documents that should be shelved. Much as he had handled the warnings from Leo Szilard and Eugene Wigner, he dismissed this, too.

Samuel Allison, a University of Chicago physicist who had recently joined the Briggs subcommittee, first heard of bombs in the presentation Oliphant made to them. As he remembered, "I thought we were making a power source for submarines." He was astonished by the prospect of a uranium bomb. Distressed and angry at the lack of attention to the MAUD report, Oliphant telephoned Ernest Lawrence, offering to come to California in order to present him with the committee's findings. In early September, he briefed Lawrence, who was so impressed by what he heard that he asked Bush and Harvard's president James Conant to meet Oliphant in Washington. Lawrence then called Compton, arranging to see him later that month in Chicago.

Finally Oliphant had found someone who shared his trepidation and grasped the enormous implications of what he was imparting. By contacting Lawrence, he had chosen well. Lawrence, energetic and entrepreneurial, had won the 1939 Nobel Prize in Physics for his development of the cyclotron. He was at the time the United States' most eminent nuclear physicist, someone Bush and Conant could not ignore. Even after meeting Oliphant, Bush and Conant remained skeptical of MAUD's conclusions, but thought they were persuasive enough to warrant ordering a new National Academy of Sciences (NAS) review. Compton was once again its head.

Up to this point, as he later admitted, Compton had been very cautious in involving Fermi because of security concerns. Fermi was not an American citizen. But Samuel Allison, Compton's Chicago colleague and protégé, told him that Fermi was absolutely the man he needed to speak to first. When Compton had asked Allison who could give him a reliable estimate of how much U-235 would be needed for a bomb, Allison's answer had come quickly, "No one can answer that question as well as Enrico Fermi."

Shortly afterward, in Fermi's Columbia office, Compton asked him the question. Fermi promptly went to the blackboard and worked out, in Compton's own words, "simply and directly, the equations from which could be calculated the critical size of a chain-reacting sphere." Fermi then proceeded to estimate how much U-235 would be needed for a bomb. It was a more conservative number than that in the MAUD report, but an obtainable one.

From New York, Compton—duly impressed by Fermi—traveled to Princeton to confer with Wigner about the relative merits of fast-neutron-induced versus slow-neutron-induced fission. After the technical discussion was over, Compton remembered Wigner urging him, "almost with tears, to help get the atomic program rolling. His lively fear that the Nazis would make the bomb first was the more impressive because from his life in Europe he knew them so well." Wigner has said that getting the U.S. government to see the value of fission had "felt like swimming in syrup." He continued to hope that America would see the risks it was running before it was too late.

Wigner's reaction was in sharp contrast to Fermi's unemotional presentation. As Szilard wrote, "Even by the middle of 1942 Fermi thought that our work had no bearing on the war and that those who thought so were sadly mistaken." Fermi had consistently refused to be an alarmist, acting only when the evidence was overwhelming. A conservative by nature, he did not like to jump to conclusions in matters of science or of politics. In many ways, this increased the weightiness of Fermi's pronouncements.

Compton was utterly convinced of the need to go forward by his own estimates and by all the arguments he had heard. The report he wrote urged undertaking the U-235 project at maximum speed. It did not mention the work on plutonium or on the pile, but Compton was already persuaded that it too was worth pursuing. He delivered the committee's report to Bush on November 7, 1941. From that point onward, things moved quickly. The sleeping giant was finally stirring.

On November 27, with MAUD's findings officially transmitted to

America and the NAS committee's report in hand, Bush went to see the president to obtain the go-ahead on the nuclear fission project. Roosevelt quickly agreed, endorsing full exchange with Britain on technical matters. The president did, however, insist that policy matters regarding the use of the project's results be restricted to a five-person group: Bush and Conant, the vice president, the secretary of war, and the army chief of staff.

Having received presidential approval and assurance of funding, Bush set about finding the best way to produce weapons based on nuclear fission. He had already begun to lay the groundwork for this project by creating yet another organization that he would head: the Office of Scientific Research and Development (OSRD). It had what the NDRC lacked, the authority to plan and develop large engineering projects. The atom bomb would eventually become by far the largest such venture the OSRD undertook.

On December 6, only nine days after his meeting with President Roosevelt, Bush summoned the uranium research leaders to Washington to tell them how he planned to organize the building of a fission bomb. Conant was appointed head of OSRD's Section One (S-1), which included key leaders of groups attempting to separate U-235 from U-238. Compton was placed in charge of bomb development. Briggs joined S-1 and the Committee on Uranium ceased to exist.

After the meeting, during the course of a lunch with Bush and Conant at Washington's Cosmos Club, Compton pressed them about a topic not mentioned in the meeting: the building of a plutonium bomb. He had already asked Glenn Seaborg to come to Chicago to brief him about the feasibility of chemically extracting plutonium produced in a pile. Seaborg had been confident it could be done. Conant, a chemist, thought Seaborg was overconfident. And Bush, the engineer, was skeptical about being able to overcome the engineering problems that would surface in trying to create large self-sustaining chain reactions. Compton was nevertheless given the authority to proceed on the plutonium project.

Everything changed radically the next day. The Japanese attacked Pearl Harbor on December 7, which President Roosevelt would call "a day which will live in infamy." On December 8, the United States Congress declared war on Japan. On the ninth, Germany and Italy declared war on the United States. America thus abruptly came to terms with its vulnerability. In a matter of minutes most of the Pacific fleet and the American air defense force stationed in Hawaii had been destroyed, and more than two thousand Americans killed. It had come as a complete surprise, with no warning. The sleeping giant had been rudely awakened.

This surfaced fears in the American scientific enterprise of a German atom bomb, once the nightmare of only a few refugee physicists. Pearl Harbor would pale by comparison to the damage such a bomb would wreak. There could be no option other than a radical acceleration of the fission research program. It must have one aim in mind: a bomb. Bush and Conant apprised Compton that he had two weeks to formulate a research schedule leading to its development.

An exhausted Compton met with them ten days later, on December 18. The timetable he proposed was that the United States should produce a chain reaction by the beginning of October 1942, a pilot plant for plutonium production by the end of 1943, and enough plutonium for one or more bombs by the end of 1944. The tasks were huge and the schedule breathtaking. Thanks to the potent and persuasive threesome of Compton, Bush, and Conant, America was poised to proceed. Within a space of a year, top-level governmental committees had been appointed, merged, and expanded in their responsibilities. They laid the infrastructure of an action agenda.

In that same December 18 meeting, Compton suggested combining fission research being conducted at Columbia, Princeton, Chicago, and Berkeley under a single umbrella. The others agreed to the plan. In January, Compton met with leaders from the four university groups. He told them that with the four scattered across the United States, it would be too complicated to coordinate efforts and to enforce secrecy. Accordingly he asked key academic scientists to meet with him in Chicago to select a single site.

On the agreed-upon day, Compton, sick with the flu, held the meeting—in his bedroom rather than his office. One of the most strategic decisions of America's nuclear future was being reached bedside, amid throat lozenges. His illness did not deter Compton from being autocratic when needed; he announced Chicago as the selected site. It would be best: the university had agreed to support the mission, and Chicago's central location made it easily accessible from both the East and West Coasts. The cover name for the center was the Metallurgical Laboratory, or more simply Met Lab.

Once the site selection meeting was over and the others had left, Compton still had to make one telephone call. That was to Fermi. Without him he knew the project could not be pursued. Fermi, the go-to physicist, Fermi, the brilliant solver of problems, Fermi, the congenial team builder, was key. Compton was also sensitive to Fermi's general reluctance to enter into anything with political overtones.

But a war was on. When he and Compton spoke, Fermi expressed some hesitation: it meant uprooting his family, leaving his lab, and abandoning his dedication to the purity of science. There was no doubt, however, in Fermi's mind. He knew Mussolini and Hitler had to be defeated. It was not only his conviction in this regard, but also his identification as an American. He was grateful that the country had welcomed him and his family; he would do what was needed for America. Fermi answered Compton in the affirmative.

25

CHICAGO BOUND

Fermi moved to Chicago at the end of April, but Laura stayed behind in Leonia until the end of June so that the children could finish the school year. Having to move again was hard for her husband, but it was even harder for her. After two years in Leonia, Laura was just beginning to feel settled. The American dream of a house in the suburbs and good schools for Giulio and Nella had become a reality.

The dream continued with the acquisition of a second car. Laura, during a visit from Hans Bethe's wife, Rose, had bought it as an essential component of suburban living. One can only imagine the car salesman dealing with two beautiful women, one with an Italian accent and the other with a German one. Little did he know they had fled respective Fascist regimes and were the wives of Nobel Prize winners, Bethe's 1967 prize yet to come. The very capable Rose negotiated smartly, and to Laura's delight there were unexpected dollars left over. The women had gotten such a good deal that Laura bought a new washing machine, a purchase she proudly announced to Enrico.

As Laura had hoped, the children seemed happy in school. Fermi described them in a glowing letter on April 5, 1941, to Amaldi as "speaking with equal ease Italian and English and being hard to distinguish from the other children in the neighborhood." As for Laura, nicknamed Lalla since childhood, he wrote, "Lalla has by now become one hundred percent American in both her habits and way of thinking." These somewhat exaggerated observations were indicative of Enrico's enthusiasm for his new life.

More candidly, Laura and Enrico carried touches of the paranoia commonly felt by those labeled as enemy aliens. They burned Nella's old second-grade reader because it was full of photographs of Mussolini. More distressing, Laura and Enrico learned from Leonia neighbors that five-year-old Giulio, apparently trying to show off, had been overheard chanting that "he wished Hitler and Mussolini would win the war." Some severe family reprimands made sure he never said that again.

Despite their comfortable American existence, Laura and Enrico were preoccupied with the war and its outcomes, their conversations repeatedly turning back to family and friends left behind. Stories about war horrors were reaching them. Their native country was being destroyed, physically and spiritually. After a run of Axis victories, however, Laura and Enrico felt a ray of hope in hearing that Germany's invasion of the Soviet Union was floundering. Launched in June 1941, it had not gone according to Hitler's plan, and the German army was stalled at the gates of Moscow, just as Napoleon's army had been more than a hundred years earlier.

Pearl Harbor came as a shock for the Fermis, as it did for all Americans, but they realized that the United States' entry into the war increased the Allies' chances of defeating Hitler and Mussolini. This made Laura happier than she otherwise might have been about moving to Chicago. She didn't know exactly what Enrico was being called to do, but was keenly aware that he would not have agreed to the dislocation had it not been important for the war effort.

While the children attended summer camp, Laura found a furnished house in Chicago near the university. It was available for rent because the owner was moving to Washington for war work. Before signing the lease, their landlord had to dismantle the short-wave reception of his radio set. As enemy aliens, the Fermis were not allowed to possess such a device. The owner also informed the two young tenants

of the house's third-floor apartment that they had to find lodging elsewhere. They were Japanese and the FBI had advised him that it would
be inadvisable to have Italians and Japanese living in the same house.

As an enemy alien, Fermi had restrictions placed on him. Most of
them, such as not being able to have a short-wave radio, a camera,
or binoculars, were minor inconveniences, but Fermi was seriously
annoyed by the constraints on his travel. Fermi shuttled between Chicago and New York for the first three months after the call from Compton. He needed special permission every time he left New York, and he
was unable to ask for a blanket exemption because of the project's
secrecy. Further miffed at having his Chicago mail opened before it
was delivered, Fermi protested to authorities. Claiming it had been a
mistake, they nevertheless continued doing so until he objected more
strenuously.

Even as Fermi was entrusted with ever more secret missions, the
bureaucracies responsible for enforcing security were slow to realize
his importance to the war mission. The military had performed a background check of him in August 1940. It concluded by saying, "His
associates like him personally and greatly admire his intellectual ability. He is undoubtedly a Fascist. It is suggested that before employing
him on matters of a secret nature, a much more careful investigation
be made. Employment of this person on secret work is not recommended." Luckily for the United States, that recommendation was not
followed.

Though Fermi was not specifically aware of the army's background
check, he would not have been ruffled to learn of it—but would have
been disturbed by its vote of no confidence. Fortunately he remained
oblivious of this fact and gave his all to the task at hand. Fermi actually had little trouble dealing with bureaucratic matters as long as they
did not interfere with his work. He had always been willing to go along
with government decrees that did not impose restrictions on his work
or on the well-being of his family. This had even been true under the
reign of Mussolini until circumstances changed.

Fermi's main objective remained fixed: to achieve a self-sustaining
chain reaction. In Chicago, after seeing that changing shapes and
arrangements of successive piles did not yield the anticipated improve-

ments, Fermi became suspicious that the uranium, central to their efforts, harbored impurities just as the graphite once had. Using the idiomatic English that his assistant Anderson was teaching him, or perhaps the somewhat mangled version he was retaining, Fermi stated to his co-workers that the uranium delivered to the lab "was filthy with dirt" and that he had been "swindled by a slick sales talk."

At the end of July 1942, thanks to a shipment of better uranium, the ninth pile built in Chicago reached the desired goal of having k, the key neutron reproduction factor, greater than one. Two more piles were built in August to ensure everything was working properly. The stage was set for building the really big pile that would decisively show if a self-sustaining critical reaction could be obtained.

The style of work Fermi adopted in Chicago was not one he was altogether pleased with. He liked to be the head of a small group, as at Columbia and before that in Via Panisperna, where he could be involved in every aspect of an experiment. But as head of Met Lab's Physics Division, he had an overwhelming amount to do. He needed to direct everybody, for he was recognized as the ultimate expert on all the experiment's facets. Fermi lamented to Segrè that his administrative role was leading him to "doing physics by the telephone." But there was no time to waste complaining. Fermi was painfully aware that victory in the war could depend on the Laboratory's outcomes.

The situation at the Met Lab was complicated because all phases of the bomb-building project took place simultaneously; each challenge was linked to the others. One could not wait until k greater than one had been reached before beginning to plan for a self-sustaining critical reaction. And one could not wait for that experiment to be completed before setting in motion plans for a much bigger pile. This pile would be called a nuclear reactor. And chemists were hard at work studying how to extract the plutonium they believed would be produced in those reactors.

As the project grew, it came to include all sorts of scientists and even a few medical personnel. According to Laura, the only people not at the Metallurgical Laboratory were, in fact, metallurgists. Of the dozens at the lab, some were Chicago personnel switching to war-time work, while others, such as Fermi, Seaborg, Szilard, and Wigner, were recruited by Compton. They in turn brought co-workers with them. Needing to find room for all of them, Compton persuaded the mathematics department to move out of Eckhart Hall, the three-story Gothic-style building that it shared with the physics department. He then had locked doors installed to guard the Met Lab's secrecy.

The individuals working on the pile behind those locked doors were initially all men. That changed in August 1942 when an attractive and athletic twenty-two-year-old woman joined them. Leona Woods, born on an Illinois farm, had earned her A.B. at the University of Chicago at eighteen and was finishing her Ph.D., working in a basement office on the other side of one of those locked doors. Herbert Anderson started talking to her and realized she had technical skills the group needed. He invited her to move to the Met Lab's side of that door. He also introduced her to his roommate, John Marshall, a young physicist working with Szilard and Fermi. She married Marshall the following year.

Woods soon met Anderson's boss and collaborator, Fermi. She described the impact Fermi had on her in a book written twenty-five years later. It begins, "Perhaps the most influential person in my life was Enrico Fermi, not only scientifically but also philosophically. He set the example of how best to deal with other people, how to anticipate change, how to put up with the ambient indignities and humiliations of the world and how to cope with the inevitable spiritual charges of taxes and death." Enrico and Laura would become her close friends.

Woods worked closely with Fermi during the war and after it. His presence was a pleasure. His qualities of "gaiety and informality . . . made it easy for the young members of the laboratory to become acquainted with him. He was an amazingly comfortable companion, rarely impatient, usually calm and mildly amused." Woods's observations echo those of all of Fermi's colleagues.

The company of young people, whether talking about science or indulging in athletic activities, was something Fermi greatly enjoyed. Leona Woods introduced him to swimming at Lake Michigan off the Fifty-Fifth Street breakwater near the university campus. This was regularly pursued at 5 p.m. in the summer and early fall; a few other young physicists often joined in. Harold Agnew, one of those Chicago swimming companions, described what it was like:

> I had been a varsity swimmer in high school so thought I was pretty good. But after fifteen minutes in the choppy, cold water of Lake Michigan, I was falling behind. Fermi, who swam with what I would call a "dog paddle" style, swam back to me and asked if I was OK . . . I barely made it to the other side of the bay and with difficulty climbed up the sea wall and sat down. Fermi said, "Meet you back where we started" and plunged back in and swam back to our starting point. I had difficulty just walking back.

Fermi, never a stylish athlete, possessed a degree of endurance and persistence that never ceased to amaze his companions.

Going to the Fermis' house for dinner would often follow. There, as Woods remembered, "Laura fed us supper after supper." The meals were relatively simple. The Fermis were not gourmets. Though he drank some wine with dinner, Fermi would add water to it first. Laura's natural grace and refinement made the dinners elegant in their own way.

Starting in that summer of 1942, the Met Lab had begun reorganizing. So far it had been a research project under the aegis of the OSRD; the pure research phase was about to come to an end. Full-scale efforts would need to be mounted if the United States was to commit itself to developing a bomb that made use of nuclear fission. There were two equally promising roads. One type of bomb would use plutonium and the other U-235. In both cases installations capable of extracting the materials for the bombs would need to be built. Compton was recommending and Conant was agreeing to go forward on both fronts. The three different methods being proposed for separating U-235 from U-238 should all be pursued because there was no way of knowing in advance which of them would be most effective. Though this would entail considerable expense, it might make the difference between victory and defeat in the war.

Conant wrote to Bush on May 14, 1942, that if time not money was to be the deciding factor in achieving nuclear weapons, going forward on all fronts might require "the commitment of perhaps $500,000,000 and quite a mess of machinery." Bush envisioned collaboration between the OSRD and the U.S. Army's Corps of Engineers. President Roosevelt agreed; the Corps was notified and in June they selected the Boston firm of Stone & Webster as principal contractors to build a structure to house the pile. Meanwhile Compton looked for a site where the pile would be erected. It had to be close enough to the city for easy access but isolated enough for both safety and secrecy. Compton found what he was looking for twenty miles southwest of Chicago, in the Argonne Forest, so named after the final big battle of World War I.

Concerned that the project was not being given sufficient priority, Bush wanted to accelerate necessary procurements for the pile and plans for subsequent bomb development. The response he got back from the army was that they would appoint a Corps of Engineers officer with a reputation for getting things accomplished and give him overall control of the mission.

In September 1942 they chose Colonel Leslie Groves, deputy chief of construction for the U.S. Army, the man who had spent the previous year building the Pentagon. There was a potential issue. Groves had overcome innumerable obstacles during the Pentagon construction. He was tired of building and wanted to be posted overseas in command of troops. However, when the Corps of Engineers commanding officer told him, "If you do the job right, it will win the war," Groves accepted. The mission's significance was underscored when Groves was told he would be promoted to brigadier general. He decided he would not take charge of the mission until the promotion had come through, for, as he put it, "I felt there might be some problems in dealing with the many academic scientists involved in the project, and I felt that my position would be stronger if they thought of me from the beginning as a general instead of as a promoted colonel."

Almost everybody who worked for the controlling, demanding, politically conservative Groves described him as abrasive, harsh, and sarcastic. The terms "son of a bitch" and "bastard" run steadily through their recollections, but they all agreed he was tireless, energetic, and effective. Living up to his reputation, Groves immediately set in motion the purchase of an enormous stockpile of uranium stored in New York by Belgium's Union Minière Company. He also laid the plan for a plant in Tennessee to separate the U-235 isotope, and convinced Delaware's DuPont chemical company to take charge of the Chicago piles as a sub-contractor to Stone & Webster.

While this was taking place, Fermi was beginning to construct the final experimental pile, the one that would sustain a critical reaction. The number of neutrons that were produced would continue to increase unless, or until, neutron-absorbing control rods were inserted to shut it down. The moment of truth for the Met Lab was rapidly approaching.

26

CRITICAL PILE (CP-1)

At this juncture, everyone at the Met Lab knew that the success of the project and perhaps the fate of the world hinged on building a large pile. Assuming the Germans were as far advanced as many thought, the scientists would have to push forward as quickly as possible. Panic that Nazis might be the sole possessors of such a weapon was leading to sleepless nights.

The Met Lab was shrouded in secrecy. Everyone working there was voluntarily committed to self-imposed censorship. Information about CP-1, standing for Critical Pile 1, stayed within the confines of the lab. The security measures, later so central to building the bomb, were still nascent. Clearance for CP-1 scientists relied strongly on peer judgment, although the army was beginning to institute formal clearance measures.

While it was understood that scientists were not to speak of what they were doing even to their spouses, Arthur Compton made an exception. As he wrote in his memoir, "I have always been one of those who must talk over important problems with his wife . . . I explained that if they cleared me it would be necessary for them to clear my wife as well, for it would be unrealistic to suppose that I would keep from her matters that strongly affected me." What he previously described as an "atmosphere of mutual confidence " extended for him beyond the lab into his home. At the end of the war, when other Met Lab wives discovered this exemption, they were none too pleased.

In spite of Laura's scientific background and their loving relationship, the Fermis never discussed the sensitive research conducted at the Metallurgical Laboratory. While one Chicago colleague interpreted this as Laura's "lack of interest in the physical world," it is much more likely she intuited the secrecy of the undertaking. Laura would not have pushed Enrico to compromise that secrecy.

In the past, Fermi had explained physics to Laura, along with anyone else willing to listen. He was an inveterate teacher. This trait was applied to laying the groundwork for CP-1 by making everyone appreciate the significance of what they were doing and to feel they were part of a team. In March, he delivered a general set of lectures about neutron physics to the Met Lab staff as a whole. He stressed CP-1's consequential nature again in September, giving more specific talks to those working directly on the pile, explaining exactly how and why it was being assembled and how it was expected to function. Doing so was a challenge because the talks had to be technical enough to satisfy physicists and accessible enough for students helping with construction. According to Herb Anderson, the talks were "fresh, clear and convincing; they showed Fermi's wisdom, his knowledge, his complete suitability for the job at hand."

The talks even had a certain amount of wry humor. Leona Woods, who was taking notes for later use, edited out some of the American slang as well as some of the analogies Fermi used. But she did leave in Fermi's comment on the key neutron reproduction factor k. If it was less than one, the reaction would not be self-sustaining. If it became much, much greater than one, Fermi's advice, spoken with a broad smile on his face, was to "run quick-like behind a big hill many miles away."

The CP-1 members began their secret work over the summer and into the early fall. When the members broke for lunch, they often headed to the university's beautiful Main Quadrangle to discuss technical issues of the project. It gave them a chance to breathe fresh air.

The Quad, framed by handsome Gothic buildings, was a large open area, crisscrossed by paths among majestic honey locust trees. With students busily rushing to and from classes, the Met team was sure that no one took notice of them and that security measures were not being breached. They spoke to one another softly, but intensely.

The task was sufficiently outrageous that the joke among participating scientists reportedly was "If people could see what we're doing with a million and a half of their dollars, they'd think we are crazy. If they knew why we are doing it, they'd know we are." Those involved were aware of the project's promise as well as its dangers, particularly the hazards of radiation tragically experienced in the past by other scientists, most famously by Marie Curie. The Met Lab included biophysicists and radiologists who focused on the physiological effects of ionizing radiations, there for research and for safety advice.

The inevitable first step in building a pile, whether small or large, was to prepare the graphite bricks that would serve as neutron moderators for the uranium pellets embedded in them. The key to the success of the effort was the quantity as well as the quality of uranium and graphite, making both the accumulation and the purity of the materials major factors. As Anderson and Walter Zinn—another physicist who had transferred with Fermi from Columbia—proceeded with their experiments under Fermi's supervision, they were constantly coping with these issues. In regard to uranium, they got a major boost from another Met team member, the chemist Frank Spedding, who enlisted his colleagues at Iowa State University to produce pure samples.

The process of building the bricks was laborious and slow. Zinn took charge of the incoming graphite, cutting the bars to the standard size needed for CP-1, then smoothing them and finally drilling two rounded holes in the 25 percent of them that would hold five-pound uranium lumps. Each brick weighed close to twenty pounds. Zinn and his crew cut and smoothed 45,000 graphite bricks and drilled 19,000 holes with shaped bottoms, each having a 3¼" diameter. It was an amazing feat of ingenuity and tenacity.

The crew that worked on the 45,000 bricks was a motley one. In addition to half a dozen young physicists and a carpenter, there were approximately thirty local boys hired from a Chicago neighborhood known as Back of the Yards, situated behind the legendary stock and meat packing yards. The impoverished area had been immortalized in Upton Sinclair's book *The Jungle* (1906); at the time of CP-1, the neighborhood was predominantly Slavic, with a large population of Poles. It became a fertile CP-1 recruiting ground for finding strong kids ready for heavy-duty lifting and drilling. One of the young physicists remembered, "These tough kids used for the production work had quit school to earn money while waiting to be drafted. These kids created a real set of challenges as some of them had a negative commitment to work." Despite this "negative commitment," the work progressed.

Many of the scientists working on the pile had already experienced being covered with graphite dust while at Columbia. They had gone home, as Fermi put it, "looking like coal miners," but that was nothing compared to the situation they experienced in Chicago. Al Wattenberg, a Columbia graduate student of Fermi's, described the omnipresent dust as a great equalizer:

> The graphite machining produced graphite dust all over the place. We breathed it, slipped on it, and it oozed out of our pores even after we washed and showered. Everyone dressed for this work in coveralls and a young professor could not be distinguished from the Back of the Yards kids.

Worried about what the work force was breathing in, Met Lab's medically trained scientists urged them to wear gas masks. The suggestion was declined with particular vehemence by the prevalence of habitual smokers. While the medical experts were defeated on this front, there were still concerns about pervasive uranium oxide dust. As a precautionary gesture, a cage of mice was placed in the area where the pile was being assembled. In this novel version of a canary in a coal mine, researchers were reassured to see that the mice seemed to be surviving nicely.

Since the end of the summer of 1942, Fermi had been sure the pile would work. It was just a question of ensuring the materials were pure enough and the engineering was done right. That confidence spread to those around him. As Anderson said, describing his relation to Fermi at the time, "It was a privilege and a thrilling experience to be associated with him in those days." By the beginning of November, Fermi was convinced they would very soon have enough high-quality graphite and uranium for the grand experiment to take place. The final pile was ready to be built. But in the meantime workers at Stone & Webster had gone on strike. The building designated to hold the final pile was not ready.

This was not the first time there had been a conflict between the physicists and Stone & Webster. A briefing earlier in the fall by the company's engineers left the assembled Met Lab staff shocked by the ignorance they displayed. Afterward, Volney Wilson, a young physicist in charge of pile instrumentation, had called a lab meeting to protest the company's involvement in the venture. What followed bordered on the surreal. Compton was in charge of the meeting, and this deeply religious son of a minister began by reading a passage from the King James Bible, Judges 7:5–7, in which the Lord directs Gideon to select good men to fight the Midianites by observing how they drink water. Then Compton sat down. A long silence followed, during which the gathering pondered how this could possibly be relevant. No one could figure it out. After a while, Wilson and others again spoke about Stone & Webster's perceived incompetence. Their discomfort was warranted.

But what could the Met Lab do now? There was no telling when the strike would be over, and who else could construct housing for the pile in the Argonne Forest? Forty-five thousand graphite bricks could be readied and significant amounts of uranium collected by the time to build the pile. The scientists could complete their tasks in this regard. But the pile would still need a place where it could be housed. With the lack of a building, the pile was homeless.

Fermi approached Compton with an unusual proposal: the physicists would build the pile themselves. They would do so right on the campus and they would have it completed no later than early December. Preparations for the experiment had been carried out in the squash courts under the stands of Stagg Field, the University of Chicago's football field. Fermi suggested the big pile be placed inside the largest of those courts, one used for doubles matches. Having redone his calculations and taking into account the purity of the materials they were receiving, he had concluded there would be enough room there. The site met the demands for space and easy access and was relatively secluded, the university having abandoned college football and the stadium a few years earlier.

But building the pile in the heart of the campus, only a few miles from downtown Chicago, posed special risks. What if it couldn't be turned off after it went critical? The ensuing meltdown would spew radioactive material into a major metropolitan area, not to speak of a campus population and the surrounding Hyde Park neighborhood. Fermi assured Compton and Groves this doomsday scenario wouldn't happen. Was there a chance he was wrong? There was great faith in Fermi, but Compton—known as a cautious man—had to ask himself if there was even a remote possibility of disaster.

Compton was also forced to anguish over a difficult question: should he or should he not notify Robert Hutchins, the charismatic and innovative forty-three-year-old president of the University of Chicago? After all, CP-1's existence at the university had been decided in good faith. Compton and Fermi had met with Hutchins to ask permission for this project vital to the war effort. Hutchins had readily understood and assured them of complete cooperation.

Despite their mutual respect, Compton finally decided that asking Hutchins would be tantamount to killing the project. Hutchins, a former professor and then dean of the Yale Law School, would almost certainly deny permission on legal as well as safety grounds. Compton concluded that the experiment was crucial. He could not risk the prospect of having the university's president forbid it. But if something went askew, Compton was aware of who would be held responsible. His name would go down in ignominy.

The decision to build the pile in the squash court was made on Saturday, November 14; construction began the following Monday. Work proceeded continuously in twelve-hour shifts, nothing new for the physicists since they had already been working ninety-hour weeks for more than a month.

Zinn headed the day shift and Anderson the night one. The first thing they did was to set down in the squash court a giant cloth balloon designed to enfold the pile. Because nitrogen absorbs neutrons, the cloth had been ordered in case it became necessary to evacuate air from the structure. This extra precaution turned out not to be needed, but the group was trying to anticipate all contingencies. Ordering it had been Anderson's idea. He admitted to receiving some curious stares from the Goodyear Company when he had done so. Experienced in building balloons, they had wondered why somebody wanted one shaped like a cube. But during a war one didn't ask too many questions about something the army was ordering.

They started to build. The squash court's interior quickly became a beehive of nonstop activity. The Back of the Yards boys who had helped shape the graphite bricks now were asked to move them. After a little over two weeks, wood scaffolding to hold the pile was in place and a portable elevator had been installed to lift the bricks and place them at designated levels.

There were no blueprints for assembling a pile or, as Woods jokingly remarked, no blackprints, a reference to the pervasive graphite dust. Fermi's technical ingenuity and improvisational skills were fully employed. The construction depended on the radiation counts they measured as they proceeded and these in turn depended on the purity of the materials. Some of the graphite was higher quality and some lower. Some of the uranium was metal, the preferred form, and some was pressed uranium oxide powder. A record was kept of each brick's location and its fabrication in case later adjustments were needed. The shape of the pile was not predetermined. Fermi had originally thought a sphere would be best. Ultimately it was egg-shaped, lying on its side, twenty-five feet across at its widest and twenty feet high, all firmly ensconced in scaffolding.

Anderson and Zinn met every day with Fermi, at which time the three of them would examine what was available and decide the optimal location for the next layers of bricks. Slots were carefully lined up so that the all-important control rods could be inserted. These thirteen-foot rods were made by nailing a sheet of cadmium to a strip of wood. Cadmium was the most potent neutron absorber available.

As material was being delivered to build the pile, negotiations were going on with the giant DuPont chemical company for the production of plutonium on an industrial scale. The company would be in charge of operating a vast enterprise at a site yet to be selected. The project, eventually built in Hanford, Washington, was so crucial that one couldn't wait for the results of CP-1; the experiment's success was assumed. Time was of the essence. On the eighteenth of November, DuPont expressed their willingness to take on the job and selected a team of their scientists and managers to visit the Met Lab for a detailed briefing. A date of December 2 was set for the meeting.

In the meantime, DuPont asked for a series of reports to explain the nature and scope of the project: these had to be written in language their managers could understand. Fermi wrote one about the feasibility of a chain reaction. In it he addressed three main questions: "Will the reaction be self-sustaining? Will the reaction be thermally stable? Will the reaction be controllable?" His answers were "yes," "probably yes," and "yes." In each case he explained them in simple terms. This was enough for DuPont.

By the twentieth of November, fifteen layers of the pile had been laid. From now on its neutron activity was measured each day at a central point of the pile with all the control rods pulled out: appropriate precautions for quickly reinserting them were always taken. The rods were held in place at other times with a simple hasp and padlock mechanism. Only Anderson and Zinn had keys to the locks.

Fermi had originally thought more than seventy layers might be needed, but the pile was working better than expected; it seemed that fewer than sixty would be sufficient. At the end of November, Fermi predicted that the pile would go critical once the fifty-seventh layer had been placed and the control rods taken out. One control rod was operated automatically, another one would be regulated manually from the squash court floor at Fermi's command, and a third was connected to the balcony at one end of the squash court.

That fifty-seventh layer was placed during the night of December 1. After taking a neutron count with all the rods but one removed, Anderson saw that Fermi had been right. He ordered all the rods reinserted, put the locks on, and went off to get some sleep.

The pile had a crude appearance, consisting of a stack of black bricks and wooden timbers. This was a primitive precursor to sleek modern-day nuclear reactors built far from urban centers and having extensive radiation shielding, elaborate cooling systems, and internal control rooms.

It had been an astonishing effort. Because the squash courts were unheated, it had meant working around the clock in freezing temperatures. A small crew had moved almost a million pounds of graphite bricks. The bricks, with holes drilled into them, held almost a hundred thousand pounds of uranium. Running through the pile, the rods were its only moving parts.

The most unbelievable feature was that all this had been accomplished in fifteen days, a miracle of planning and cooperation. It was a sterling example of how academia, government, and industry could work effectively together and even keep things secret in the process. There were no leaks of any kind, neither of information nor of radiation. It was inspiring how Americans and recent emigrants from fascist countries cooperated. The top secret project in the United States owed much to these refugees. And notably, Fermi's classification as an enemy alien had been lifted less than two months earlier than December 2, 1942.

27

THE DAY THE ATOMIC AGE
WAS BORN

Fermi's steady hand did not waver throughout the construction process of the pile, his colleagues agreeing that he seemed completely self-confident. As leader of the team, he threw himself into every phase of the preparations, never pulling rank or displaying a modicum of conceit. His precision, down to forecasting exactly when the last brick had to be placed, was a source of wonder.

On the night of December 1, Fermi slept well, unperturbed by doubts: he was certain that the following day would be successful. Early the next morning Fermi made his way to the court, stopping to pick up Leona Woods, who lived nearby. They trudged together slowly through the snow in the subzero weather. Chicago, the Windy City, lived up to its reputation as raw winds blew off Lake Michigan. They walked in silence, the freezing air stilling their voices. Thoughts of the day ahead preoccupied them.

At the squash court they met Walter Zinn; they checked with him that all the equipment was in place. And then a still sleepy Herbert Anderson appeared. He had had only a few hours of rest after coming off the night shift. The pile, carefully minded around the clock by either Zinn or Anderson, was never left alone. Anderson was hungry, so

Fermi and he went with Woods for a quick breakfast at the apartment she shared with her sister. She hurriedly made pancakes, and then the three returned for final preparations.

The squash court had a balcony at one end, originally intended for spectators to look down at an ongoing game. The December 2 event was no game. The view from the balcony that morning was not of players moving nimbly about but of a large immobile black egg-shaped object encased in a nest of wooden timbers. Scientists, Fermi foremost among them, would replace the spectators on the balcony. On the court, a trusted lone physicist tended the egg.

The balcony held various recording devices for monitoring the neutron count, their scales adjusted to enable continuous recordings of the count rates. Underlying the setup was the assumption that the pile would go critical. In anticipation, the safety mechanisms had been installed. The thirteen-foot control rods that absorbed neutrons were all in place, ready to be pushed in or pulled out of the pile. Zinn had devised the special weighted one poised above the pile. Given the name of Zip, it was held there by an electromagnet connected to an ionization counter measuring neutron activity. If the activity level passed a certain preset safety value, Zip would be automatically released and descend into the pile.

The control rod tied with a rope to the balcony had an extra and unusual safety feature. There stood Compton's associate director, Norman Hilbery, with an ax in hand. If all else failed, he would chop the rope. CP-1 technology ran the range from the most sophisticated to the most primitive.

If all other safety measures failed, there were three buckets of cadmium sulfate, held by a safety squad, with orders to splash the pile if it went out of control. The cadmium sulfate would absorb neutrons, aborting the experiment. It would also effectively destroy the costly black egg.

Midmorning, after checking out the equipment with other team members, Fermi was ready to commence the trial. Respective roles had been carefully delineated. Fermi, the director, was giving orders from the balcony above. George Weil, another Columbia physicist who had come to Chicago with Fermi, was entrusted to be the executor of orders, manipulating the rods on the court below. Woods, the only female and the youngest person there, was in charge of calling out the counts. The role of the spectators was easy: keep quiet and continue breathing.

All the control rods, other than the one guided by Weil, were pulled out. After the safety measures were verified as functioning properly, Fermi told Weil to pull the last one halfway out. The neutron count, read off by Woods, rose quickly and then leveled off precisely as Fermi had predicted. Her voice resounded strongly over the male assemblage. Volney Wilson, with Fermi on the balcony, checked the counters once again and set them to record a higher rate.

Confident that everything was working as anticipated, Fermi directed Weil to continue pulling the last rod out, six inches at a time. Fermi checked the neutron count after each extraction to ensure it matched his calculations. Tension was mounting among the spectators after the third six-inch outward slide of the rod because they knew the pile was close to the point where a critical reaction would set in.

At 11:30 a.m. everybody jumped as a sudden boom shook the room. The Zip control rod had fallen in, effectively turning the pile off. It turned out they had simply been too cautious in setting the radiation count value at which Zip would drop. Nerves were on edge. Fermi then said, "I'm hungry. Let's go to lunch." He sensed his team needed a break. Food was always good for nerves. All the rods were put back into place and locked in by Zinn and Anderson.

After lunch they gathered again at the squash court, now joined by Compton, who had not been there in the morning. He was meeting with DuPont personnel and an external review committee appointed by General Groves. The committee had been formed to assuage DuPont's understandable doubts about agreeing to build a plutonium production enterprise based on a process not yet achieved in a laboratory. It was indicative of Groves's faith in Fermi that he proceeded with that plan before knowing that CP-1 would reach criticality.

When Compton was informed that Fermi would be conducting the momentous trial in the afternoon, he decided the review committee would have to do without him. Although by then the balcony had limited space, Compton brought one member of the DuPont staff, Crawford

Greenewalt, a dynamic forty-year-old chemical engineer who led the DuPont contingent that had come to Chicago.

The experiment resumed at two in the afternoon. Fifty people would later lay claim to having been at the CP-1 experiment but there were only around forty at the moment when the pile went critical. Each of those present was aware of the dangers of the trial but each wished to bear witness. No one was running for the hills as Fermi had previously jested.

Fermi again instructed that all the control rods but one be removed from the pile. After checking that the neutron radiation count was exactly the same as it had been earlier, he directed Weil to gradually slide out the last one, first to the halfway point, and then more slowly beyond that.

Anderson, by Fermi's side as the order was given to continue removing the rod, remembered what it had been like: "Again and again, the scale of the recorder had to be changed to accommodate the neutron intensity increasing more and more rapidly. Suddenly Fermi raised his hand and then announced, 'The pile has gone critical.'"

Anxiety grew as Fermi continued to let the pile run, the neutron count growing steadily. In his usual manner, Fermi remained calm. One minute passed, then a second and a third. After little more than a fourth minute, with the tension in the room becoming almost unbearable, Fermi ordered, "Zip in!" It was 3:53 in the afternoon. The control rod was immediately lowered. The intensity of the neutrons dropped as quickly as the stress in the room. Everyone resumed breathing.

Compton described Fermi at the crucial moment as "alert, in as full control of his experimental crew as is the captain of a ship engaged in critical action. At this moment of great achievement his face showed no signs of elation. The experiment had worked precisely as expected. Cool, collected, Fermi's mind was not dwelling on the significance of what had just been done. He was laying his plans for the next urgent stage of the work."

Bigger piles would have to be built. They would run for days or months, not minutes. Their products would need to be extracted. What had been accomplished today was only a beginning. It was a historic turning point that marked the birth of the atomic age. But there could be no letup.

Compton and Greenewalt were among the first to leave the squash court after the pile had gone critical. Greenewalt hurried back to the review committee to give them the news. Compton remembered the expression on his face: "His eyes were aglow. He had seen a miracle . . . his mind was swarming with ideas of how atomic energy could mean great things in the practical life of men and women. As an industrial engineer, war at this moment was far from his mind. Here was a source of endless power that could warm people's homes, light their lamps and turn the wheels of industry."

Other spectators who had witnessed what had just taken place had different thoughts. After the triumphant demonstration of a first chain reaction, wild cheers did not break out. The event was exciting but also sobering. It did not take long for those present to start asking what this might mean for the war effort. If they had succeeded in obtaining a self-sustaining critical reaction, wasn't it probable that the Third Reich had been able to do so as well? And what might the Germans do next— or what had they already done? There would be no time to pause. Work toward a bomb would have to proceed with even greater resolve. Fermi knew this, and so did everybody else who was there in the squash court.

It was up to Compton to relay the news to Washington. He stopped by his office after leaving the squash court and placed a phone call to James Conant. He remembered their exchange, couched cryptically in case it was being overheard. Compton began:

"Jim," I said, "you'll be interested to know that the Italian navigator has just landed in the new world."

Then half apologetically, because I had led the S-1 Committee to believe that it would be another week or more before the pile could be completed, I added, "the earth was not as large as he had estimated and he arrived at the new world sooner than he had expected."

Conant's excited response was "Is that so, were the natives friendly?" I answered, "Everyone landed safe and happy."

As darkness began to descend on Chicago that December afternoon, those in the squash court slowly drifted out. Finally, just as the last twenty or so were beginning to take off their gray lab coats, Wigner produced a straw-covered *fiasco* of Chianti and some paper cups that he had stored in a bag on the squash court floor. Fermi uncorked it and poured a sip for each of those present. Someone asked Fermi to sign the bottle's covering and it was then passed around for the others to sign.

Zinn, who had been the first to greet Fermi at the pile in the morning, was the last physicist to leave in the evening. He had wanted to double-check all the apparatus and make sure once again that all the rods were firmly locked in place. When he finally filed out, one of the guards stationed outside asked him, "What's going on, Doctor, something happen in there?"

Something had indeed happened. None of those in the squash court that afternoon forgot being there when the pile went critical. It had only generated a maximum power of half a watt, scarcely enough to light a flashlight battery. However, if that rate had been allowed to grow unchecked, it would have killed everyone in the squash court and wreaked havoc on the city of Chicago.

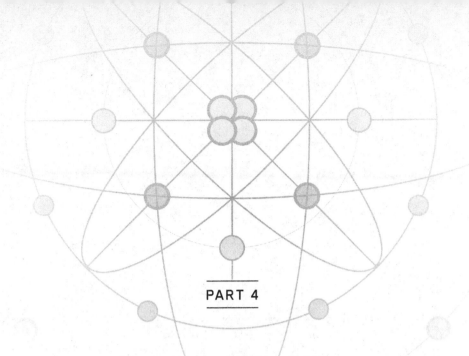

PART 4

THE ATOMIC CITY

28

THE MANHATTAN PROJECT:
A THREE-LEGGED STOOL

The first official, albeit secret, description of what happened on the Chicago squash court was straightforward. Written by Fermi with no rhetorical flourishes, it appeared in the monthly Met Lab report, an internal document circulated only among those who already knew they were part of history. Fermi stated: "The chain reacting structure has been completed on December 2 and has been in operation since then in a satisfactory way."

Twelve years later, on the day of Fermi's death, a famous American broadcaster adopted a profoundly different tone in describing the stupendous event. On the CBS evening news, Edward R. Murrow reported: "The story of the lighting of the first atomic furnace will be told as long as stories can be listened to, for it was certainly one of the most dramatic moments in the unfolding of human knowledge." Murrow then commented, "It was the good fortune of this country that Dr. Fermi found asylum in 1938. Under the present immigration laws, he might not be admissible . . . His exclusion would be shared by other immigrant founding fathers of the atomic age." With trademark astuteness, Murrow summarized Fermi's brilliance and the conflicts of the times.

For Laura Fermi, the day of December 2, 1942, was much like other days except that Enrico returned from work a little earlier than usual. She thought he had simply done so to help prepare for a large party for his co-workers. Laura's schedule had been thrown off by six-year-old Giulio, caught in a mischievous infraction. While his father was bombarding uranium with neutrons on a squash court, his son had been bombarding the neighbor's sparkling windows with dirty snowballs. Enrico walked in to the tail end of a prolonged scolding of Giulio by Laura and the child's ensuing remorse. Probably still engrossed with the criticality of the pile, he shrugged off his son's misbehavior. After dinner, Laura and Enrico quickly organized things by moving chairs, choosing records for dancing, and lighting some candles. With the winter solstice fast approaching, the Fermis wanted the festivities to be cozy and welcoming. A glass of wine or red vermouth would help.

The timing of the long-scheduled party was completely coincidental with the day's discovery. Shortly after eight the first guests, Walter Zinn and his wife, Jean, arrived. Laura was surprised to see Zinn, with a big smile on his face, shake Enrico's hand and say, "Congratulations." Laura did not have a whiff of anything exceptional happening that day. Secrecy shrouded Enrico's current pursuits and therefore conversations about it were prohibited. But still, her husband had given no indication of anything of note having taken place, had not seemed the least bit nervous the night before, and acted as though tomorrow would offer nothing out of the ordinary.

Laura's curiosity grew as the pattern of congratulations repeated itself time after time with other guests. Thinking that Leona Woods was the most likely person to tell her, Laura asked her what Enrico had done to deserve so many warm kudos. Because of security rules, Woods was at a loss for words. Flustered, and perhaps subconsciously thinking the pile might soon be used in retaliation for Pearl Harbor, she blurted out, "He sank a Japanese admiral." Laura looked askance but Herbert Anderson, Fermi's right-hand man since Columbia days, came to the rescue, saying Laura must know that "anything was possible for Enrico." Laura dismissed it as some sort of private joke and went back to entertaining guests. The mood among the partygoers was celebratory, but tempered by the magnitude and meaning of what had happened on the squash court earlier that day.

The next morning, Fermi slept a little longer than usual. Normally he was up at 4 a.m., which Fermi regarded as his form of insomnia, working at his desk at home for a few hours before heading to the lab. His routines seldom varied and colleagues would joke that they could set their watches according to Fermi's comings and goings. He was in a very good mood that Thursday. No longer having to focus on whether or not the pile would work, he could apply its results to exciting new physics experiments. As Anderson wrote, "What thrilled Fermi most about the chain reacting pile was not so much its obvious promise for atomic energy and atomic bombs which many others were now prepared to pursue, but an entirely new and unsuspected feature. It was a marvelous experimental tool." Obtaining greater intensity in neutron sources had been a continual challenge for Fermi since 1934 and now all he had to do was move a control rod in or out a little bit. With a simple gesture he could increase the neutron flux by a factor of a million. It was like waving a magic wand. Years later Fermi commented, "To operate a pile is just as easy as to keep a car running on a straight road."

By mid-February, there was an improved version of the experimental tool. CP-1, the first pile, was taken apart and rebuilt in the Argonne Forest, the site originally designated for assembling the pile. Erecting it in the squash court had been a matter of expediency but now that it had proven itself, a more permanent pile was warranted. The foremost change in design was the erection of a concrete shield surrounding it, allowing the pile to be operated at a much higher power without exposing observers to radiation. CP-2 typically ran at 100 kilowatts, more than 100,000 times as great a power as CP-1. This time it was not physicists building it, frantically hauling graphite blocks, but the firm of Stone & Webster, which had come back into government's good graces after their initial debacle.

Testing graphite and uranium samples for purity, once such an effort, was now made easy. Finding the best pile configuration went smoothly because one wall of the shield was movable, allowing access to the pile's interior. Fermi delighted in going back to basic research, "the work he enjoyed most," according to Anderson. Conditions at Argonne were pleasant: a quiet forest to walk in and enthusiastic young colleagues surrounding him. But the hiatus was brief for Fermi. The unrelenting pressure of the war effort could not be ignored, no matter how fascinating were the results of his research.

Even before the success of the pile, General Groves had been moving at lightning speed to prepare for the further development of atomic weaponry. His track record of taking on giant projects and moving them aggressively reached new heights. With the implicit backing of President Roosevelt, who had signed a secret enabling document on May 12, 1942, Groves led what would first be called the Development of Substitute Materials Program. But the title was considered too revealing, so it became the Manhattan Engineer District, with headquarters located at 270 Broadway, eighteenth floor, in New York City. In turn, it became known as the Manhattan Project, evolving into a vast research and development empire of more than thirty sites spread across the United States. Groves left no one in doubt as to who was in charge.

Within a few months' time, the Manhattan Project had bought a total of more than 140,000 acres for three atomic communities devoted to conducting secret research. Site X was near Clinton, Tennessee. Groves had approved it within days of his appointment and it was purchased on September 29, 1942. Site Y, Los Alamos, New Mexico, was next, purchased on November 25. And Site W, near Hanford, Washington, was last, on February 9, 1943.

Each site, X, Y, and W, had a distinct purpose but a shared mission: to develop a nuclear bomb. And Fermi would leave his fingerprints on all of them, to varying degrees. He was least involved in Site X, Oak Ridge, although the design of its large nuclear reactor resulted directly from Fermi's successful nuclear pile. At Site W, Hanford, Fermi's involvement was key in solving how an adequate amount of plutonium could be produced to make a bomb. At Site Y, Los Alamos, Fermi was a valued member of the team that put all the pieces together. He was the person to consult about almost any physics question. He tended to either answer it or to frame it so a route could be found to a solution.

In his scientific career, Fermi had stood out—had in fact, shone—

with his unique contributions. When he became involved with the Manhattan Project, he was one in a community of fellow geniuses with whom he worked side by side. Before, Fermi had made history with his discoveries. Now, with the Manhattan Project, he was caught in the sweep of history.

Groves made certain the Manhattan Project was given the highest priority rating, AAA, reserved by the War Production Board for "emergencies." Arguing that time, not money was crucial, he bulldozed his way through the bureaucracy. When the civilian head of the board hesitated in granting the request, Groves threatened to bring the issue directly to the president. The AAA rating was promptly bestowed.

Both Sites X and W needed to be proximal to abundant water to generate electric power for running their respective plants and for the cooling reactors. Oak Ridge was aided by the presence of the Tennessee Valley Authority (TVA) and the nearby Clinch River. Groves immediately set to building a reactor bigger than Argonne's CP-2. Fermi and his group guided the process.

The most formidable task at Site X was separating U-235 from the much more plentiful U-238 ore. Its reactor was scheduled to be up and running by the end of 1943 and was expected to produce enough plutonium for further studies and applications. But the large amounts of plutonium needed for a bomb would have to be produced at a separate location. Site X was uncomfortably close to Knoxville, Tennessee. If there was a disaster, lives would be endangered and it would compromise Oak Ridge's uranium ore plants. The two avenues toward building a bomb, U-235 and plutonium, had to be kept clear of one another.

The remoteness of Site W, Hanford, allowed space for three nuclear reactors, ten miles apart, and three chemical separation plants for extracting plutonium from the irradiated uranium. It was an enormous industrial enterprise. While CP-1 had run at half a watt and CP-2 at 100,000, the Hanford reactors began functioning at 250,000 watts and ramped up from there to 100 million watts. Each one of them was cooled by 30,000 gallons of cold Columbia River water pumped through the reactor every minute. Mesh on water intake pipes ensured that none of the precious Columbia River salmon were sucked up. A professional fish scientist was hired to monitor the health of the fish. The water was then stored in large holding basins nicknamed Queen Marys, after the ocean liner. Once radioactivity had diminished to levels the scientists regarded as safe, the water was returned to the river.

In addition to building operational plants, both Sites X and W needed to create residential communities for the projects' employees. Site X at Oak Ridge evolved into a town of seventy thousand inhabitants by the end of the war. It was designed by the upstart architectural firm of Skidmore, Owings and Merrill in Chicago, which was asked to submit in four days a master plan for housing and amenities. The group of ten architects had no idea where the city would be built or of its purpose, but they managed to get the job done. Site W at Hanford similarly developed into a community, housing fifty thousand workers.

But Oak Ridge and Hanford were only two legs of the three-legged stool. Sites X and W would produce the necessary materials for an atomic bomb. How to take those materials and forge a weapon out of them was the third leg of the stool. Site Y would be responsible for the assembly and detonation of a bomb. For that, other kinds of skills and talents would need to be tapped.

At the top of the agenda was who should lead such an effort. The second item was where Site Y should be located. On October 8, 1942, Groves met J. Robert Oppenheimer for the first time at a Berkeley luncheon hosted by the president of the University of California. The two men were a study in contrasts: both tall, but Groves burly and brawny and Oppenheimer stooped and fragile; Groves, the son of a Presbyterian army chaplain, and Oppenheimer, the son of assimilated cultured Jews; Groves gruff and outspoken, Oppenheimer charming and soft-spoken. Yet the two somehow recognized in each other impressive intellect and drive, albeit channeled in different ways.

In response to Groves's questions after the Berkeley lunch, Oppenheimer exercised characteristic charisma. He adroitly outlined for Groves his vision of a separate laboratory where experts in different fields would come together to design and test a bomb light enough to be loaded on an airplane. He added that since there were so many unknowns in its development, work on the laboratory should begin at once. Groves was struck by Oppenheimer's genius but mostly by his "overweening ambition." It was a trait to which Groves readily related.

Groves was much in favor of Oppenheimer's vision, but there were considerable obstacles. Should the laboratory be under military control? Where should it be located? Who should lead it? The director needed to have a stellar scientific reputation and administrative acumen, to be persuasive in recruiting top scientists, and adept at negotiating between them and the military.

Arthur Compton, Ernest Lawrence, and Harold Urey, all three of them Nobel Prize winners, were the obvious candidates but they were already engaged in key war-related efforts. Fermi was eliminated, too. Foreigners who obtained clearance would be allowed to work on the Manhattan Project, but the army would not tolerate having one in a leadership role. A week after meeting Oppenheimer, Groves had settled on Oppenheimer and offered him the position as director of Site Y.

Many people thought the decision strange. They knew Groves put a high premium on a director's possessing a Nobel Prize, but Oppenheimer had not achieved that honor. Moreover, Oppenheimer had never chaired a physics department or even shown much interest in experimental research. Then there were his left-wing political leanings. Because of his associations with Communists, Oppenheimer could be considered a security risk. But Groves wanted him and asserted that Oppenheimer was essential to the project.

When looked at more closely, this decision was sound. In the summer of 1942, Oppenheimer had organized a secret seminar in Berkeley during which a small number of eminent theoretical physicists attempted to determine what was necessary for building an atom bomb. The leading young theorist of nuclear physics, Hans Bethe, had been reluctant to attend, but Oppenheimer had persuaded him to come. And he also convinced Edward Teller, Felix Bloch, and others. They formed a group that Oppenheimer called his luminaries.

This elite assemblage concluded that while a nuclear weapon was feasible, building it would require more fissile material than earlier estimates and much more experimental information. They also appreciated Oppenheimer's brilliance. Groves had chosen well.

In late October 1942, after appointing Oppenheimer, Groves pondered where Site Y could be located. It needed to be away from prying eyes, have a temperate climate for outdoor testing, and be accessible from physics centers on both the West and East Coasts. Because of fears of infiltration, a further requirement was that its location be at least two hundred miles from any international boundary. The neighborhood of Albuquerque, New Mexico, seemed particularly well suited: Albuquerque was a stop on the Atchison, Topeka and Santa Fe rail line and for TWA transcontinental flights.

After a few potential sites were rejected, Groves and Oppenheimer met in New Mexico in mid-November. Oppenheimer knew that part of the world very well. He had once written a friend, "My two great loves are physics and the desert country. It's a pity they can't be combined." He now found that they could. He suggested to Groves, "If you go on up the canyon you come up on top of the mesa and there's a boys' school there which might be a usable site." Site Y, the third leg of the Manhattan Project, had been found.

29

SIGNOR FERMI BECOMES MISTER FARMER

The mesa is an enchanting spot, with a backdrop of the eleven-thousand-foot Jemez Mountains thick with ponderosa pine forests and fir. At lower elevations the fragrance of the pines mingles with that of piñon trees and sagebrush. The breathtaking view across from the mesa looks toward the even higher Sangre de Cristo range, replete with aspen trees, alpine meadows, and craggy mountains capped by snow most months of the year. The sun's rays, hitting the Sangres with a warm glow, are responsible for its name: Blood of Christ. In between the two ranges lies the wide valley of the Rio Grande, majestically winding its way through deep canyons and high mesas of every shape. The clarity of the air, the constancy of the sun, and the stark beauty of the landscape are exhilarating.

The small boys' boarding school was surrounded by old poplar trees, known in Spanish as alamo trees, and comprised the "Big House," an old lodge built out of massive hand-hewn pine trunks, a dormitory, a pond, a few smaller buildings, and a corral with stables for the boys' horses. It was situated on eight hundred acres, 7,200 feet above sea level.

Groves set the purchase in motion. The school closed on January 22, 1943, and was totally vacated by the end of February. Construction began almost immediately on barracks for housing the personnel, roads to transport material up to the mesa, fences to enclose the new laboratory, and a gatehouse that became the main entrance and exit for the secret city.

Meanwhile, Oppenheimer traveled around the country searching for scientific personnel, relentless in his pursuit, self-describing his recruiting as "absolutely unscrupulous." It wasn't easy, because so many scientists were already engaged in war work. Fermi, one of the first people he approached, was completely tied up with the experiments at Argonne that were essential in informing the building of the reactors at Oak Ridge and Hanford. He also contacted Isidor Rabi, the associate director of MIT's Radiation Laboratory. Rabi was considered indispensable to MIT's intensive radar research activities, but he convinced a few others to join Oppenheimer.

The first scientists arrived in the middle of March 1943 at Site Y. Groves had anticipated that they would differ from personnel at Sites X and W: "We were faced with the necessity of importing a group of highly talented specialists some of whom would be prima donnas, and of keeping them satisfied with their working and living conditions." Prima donnas or not, they were willing to venture forth and put other research on hold.

Oppenheimer gathered a formidable set of scientists in Los Alamos, superstars in the physics world. In the words of Bob Wilson, head of the lab's cyclotron group:

> The projected laboratory as described by Oppenheimer sounded romantic—and it was romantic. Everything to do with it was to be clothed in deepest secrecy. We were all to join the army and then disappear to a mountain—to a laboratory in New Mexico—Los Alamos. It sounded especially romantic to me for I had just finished reading *The Magic Mountain* by Thomas Mann. I almost expected to come down with tuberculosis, certainly expected to explore the philosophical significance of concepts of time and space and freedom and fascism in intense conversations with an Italian philosopher in a snowstorm. And I did, too! Not with Mann's fictional Settembrini, but rather with a real, live, breathing Enrico Fermi.

Wilson knew that it would be a while before the "real, live, breathing Enrico Fermi" would join the Los Alamos ranks, but it was clear that an enormous reservoir of talent was being recruited. It helped attract others—famous and not so famous. There was the lure of camaraderie, of empowerment to contribute to the wartime effort, and even of destiny.

In April, Oppenheimer organized an orientation meeting for the new recruits and prevailed upon both Rabi and Fermi to attend. Five introductory lectures were designed to explain the mission of the Los Alamos Scientific Laboratory (LASL) to the congregated scientists. The physicists among the approximately hundred in attendance had a pretty keen sense of what they would be hearing but others, such as chemists and metallurgists, were still wondering.

Delivered by Oppenheimer's close collaborator Robert Serber, the lectures were essentially an updated version of the Berkeley 1942 summer group discussions. The meeting notes were later condensed into a twenty-four-page summary that became the laboratory's first publication, LA-1. Known as the *Los Alamos Primer*, it was handed to new arrivals for the next two years. Its first paragraph sets out the laboratory's goal: "The object of the project is to produce a practical military weapon in the form of a bomb in which the energy is released by a fast neutron chain reaction in one or more of the materials known to show nuclear fission." One change in language took place soon afterward; at Oppenheimer's suggestion, the weapon was referred to in the laboratory as the Gadget, not as a bomb.

The tasks ahead were daunting. Fermi was impressed by the personnel's spirit. Oppenheimer remembered that after one of the lectures, Fermi said to him, in the ironic tone he sometimes adopted, "I believe your people actually want to make a bomb." If they had not thought of making a bomb beforehand, the lectures served to consolidate their resolve. Whether motivated by nationalism, fear of Hitler, or the lure of physics, the Los Alamos group pulled together. Later on, some would accuse them of making a Faustian bargain. But with Hitler still in power, reports of the mass slaughter of Jews confirmed, and Japan showing no signs of surrender, there seemed little choice.

Acting as senior consultant, Fermi returned frequently to Los

Alamos; it was anticipated he would eventually move there. But for the time being, his role at Argonne, Hanford, and Oak Ridge took precedence. Conscious of travel demands and the clandestine world of a top secret military mission, Groves made sure the premier scientists associated with the Manhattan Project were afforded adequate protection.

Groves asked Fermi not to travel by air unless it was essential. While traveling, he was to be known as Henry Farmer. And he was assigned a bodyguard whose orders were to accompany Mr. Farmer everywhere and shield him from danger. With Fermi's nom de guerre, Laura was amused to see her husband returned to the agrarian roots his family had so proudly escaped. Enrico, who did not even enjoy gardening, was back to farming, albeit in name only. And she, whose ancestors had perhaps never farmed, was suddenly the wife of one.

Laura described her trepidation upon first encountering Enrico's bodyguard, John Baudino, in Chicago early in 1943. "One evening I went to answer our doorbell and found myself face to face with a big man who entirely filled the doorway . . . In a deep voice with no harsh tones, the apparition shyly asked me to tell Dr. Fermi that he would wait for him outside." Since Fermi had always valued his independence, Laura worried about how he might adapt to having someone constantly at his side.

But Baudino, a twenty-nine-year-old recent graduate of the University of Illinois Law School, was so good-natured and agreeable that there were never any problems. When walking outside the Fermi house or on the campus, Baudino accompanied Fermi. On daily trips to Argonne, Fermi did the driving, since he didn't like having others at the wheel, and Baudino sat in the passenger seat. On trips the two of them took to Los Alamos, Oak Ridge, or Hanford, they would often play gin rummy, Baudino keeping a record of the winnings. An associate remembered, "By the time Baudino returned to civilian life, Fermi owed him several million dollars. They were both perfectly straight-faced about this debt."

Fermi's involvement with Oak Ridge was not extensive. He and Arthur Compton went there in November 1943 to observe the X-10 reactor going critical for the first time, but comparatively little was asked of him. Hanford, still in the planning stage, was another story.

Design for the Hanford plant was the Argonne Laboratory's first priority from the summer of 1943 until mid-September 1944.

The pace of work quickened with the Argonne group testing materials and designing the reactor's control rods. Leona Woods, who married her co-worker John Marshall in July 1943 and had become pregnant shortly afterward, recalled working right up to the time the baby was born. As she noted, "My work clothes—overalls and a blue denim jacket—concealed the bulge." She didn't tell Walter Zinn, who was in charge of CP-2 operations, about her pregnancy "because Zinn would probably have insisted on kicking me out of the reactor building."

Fermi did know about her pregnancy for a while, and when it became obvious that the birth was imminent, he worried about whether there was enough time to drive from Argonne to the University of Chicago's Lying-In Hospital. Preparing for all eventualities, Fermi asked Laura for instructions on delivering a baby. This did not please Leona Woods Marshall: "When he told me he was ready, it stiffened my resolution that under no circumstances would he get the chance to practice midwifery." Fortunately, Fermi's newly acquired, although limited, skill was not solicited. After a healthy delivery in the hospital, Marshall was back at work at CP-2 within a week.

Most of the work Fermi accomplished in early 1944 was carried out at Argonne, but by midyear he was making more frequent trips to Hanford. Several members of the Argonne group, including the Marshalls with their new baby, felt that moving there was necessary to oversee the reactor's workings. Activity in Los Alamos was also reaching a pivotal stage, and Fermi's presence was urgently requested there as well.

The family scheduled their move to the mesa for mid-August 1944. Fermi was eager: the scientific atmosphere there was stimulating, the challenge exciting, and the environs well suited to his love of the outdoors. There was also more than a touch of patriotic fervor. On July 11, 1944, swearing their allegiance to the United States before a Chicago District Court judge, Laura and Enrico Fermi had become United States citizens. The family, who had been labeled as enemy aliens a few years before, took the solemn oath of loyalty and swore to "absolutely and entirely renounce and abjure all allegiance and fidelity to any foreign prince, potentate, state, or sovereignty, of whom or which I have heretofore been a subject or citizen." The Fermis had by then decided their future was firmly planted in America. Living in Italy, even after the war, was not an option. Enrico's quip upon disembarking in New York in January 1939 that "We have founded the American branch of the Fermi family" was now backed by the family's official status.

A few days before the family's move, Fermi was summoned to Site W in the state of Washington. Laura would have liked Fermi by her side as she traveled with the children to this peculiar destination, but she was intrigued by a new adventure and by now reconciled to his recurrent absences. Weighing whether to stay in Chicago until Fermi got back, she had asked him when that might be. He confessed, "I haven't the faintest idea."

In any case, Chicago was hot and humid in the summer, whereas Enrico had reported that New Mexico in August was dry and pleasant. The children, Giulio, now eight, and Nella, thirteen, were smitten by images of the Wild West and the prospect of Indians, cowboys, and horseback riding. But at the same time they were reluctant to leave their Chicago home and their classes at the progressive University of Chicago Laboratory School. They would be starting a new school in Los Alamos, one that Laura had been assured would be excellent, thanks to the many highly educated physicists' wives who were eager to have meaningful employment as teachers while their husbands were off working on the Gadget.

And so Mrs. Farmer and the young Farmers boarded the train to Site Y while Mr. Farmer left for Site W. Enrico assured his family they would like Los Alamos but provided few details of what they might expect. Laura was only told that she would be met when she got off the train in Lamy, New Mexico. A drive to Santa Fe brought them to 109 East Palace Avenue, the small, inconspicuous office of Dorothy McKibbin, who inevitably served as the first point of contact for those bound for Los Alamos. McKibbin greeted the Chicago family warmly, putting them at ease within the office's protective adobe walls and arranging for their ride up to "the Hill." She also issued them the required security passes to enter and exit the gated city. She was known as the gatekeeper of Los Alamos, the ultimate secret city, its address only Post Office Box 1663, Santa Fe.

The drive from Santa Fe to Los Alamos has to count among the world's most stunning. Starting from one of America's oldest settlements, founded in the early seventeenth century by the conquistadors, one leaves quaint adobe dwellings sheltered by giant cottonwood trees and heads toward arid desert marked by tumbleweed, scruffy shrubs, arroyos, and remnants of volcanic lava from centuries ago. The road, turning into a washboard surface, skirts Indian pueblos and crosses the Rio Grande before climbing steeply and often treacherously up among ancient cliff dwellings and towering mesas. Every bend in the road offers even more dramatic views of the encircling mountains.

The magnificence of the New Mexico scenery contrasted sharply with the unattractive town of Los Alamos, hastily erected with regimented rows of green dormitories and apartments. It had all the appearances of the army base that it was. The old houses of the boys' school were an exception, its lodge and the nearby buildings reserved for the lab's scientific crème de la crème. Known as Bathtub Row because they included the luxury of bathtubs, in contrast to ordinary showers in other housing, they were premier dwellings.

Given her husband's distinction, Laura could have insisted on living on Bathtub Row, but she was content with a three-bedroom apartment in building T-186, a typical two-story four-apartment wooden structure. Admittedly, Laura was somewhat taken aback to see blankets and sheets with USED stamped on them, although relieved when she was informed the army supplies were not "used" but had been issued by the United States Engineer Detachment. It was still a far cry from the ironed linen sheets she had grown up with and that she continued to use in Chicago.

The apartment was satisfactory in spite of shoddy workmanship and minimal insulation from weather and noise, but the accommodations were made more palatable by Laura's discovery that the apartment below was occupied by a couple she had known in Rome a decade earlier, the German-born physicist Rudolf Peierls and his ebullient Russian wife, Genia. They had been in Los Alamos for a while and were delighted to take Laura under their wings. It helped that they had a daughter a year younger than Nella and a boy a year older than Giulio.

The morning after Laura and the children arrived, one of the Boys of Via Panisperna appeared. Emilio Segrè had moved to Los Alamos in 1943 and announced, in the spirit of former times, "While the Pope is not here, I'll bring your mail." Perhaps it was America's more secular atmosphere, but the title of the Pope had seldom been used once Fermi had crossed the Atlantic. Now, in such a strange and wondrous setting, it felt welcoming to Laura.

When she inquired why mail delivery by Emilio was necessary, he replied, "There is no home delivery. All the mail goes to the Tech Area and there it stops." It did more than stop there; the mail was read and censored lest information be leaked. Since correspondents were not supposed to know that the mail they were receiving had been censored, outgoing letters were deposited unsealed. If they passed inspection, the censor would seal them and send them on. If not, they were returned to the sender.

Much to Laura's amazement, Emilio, who had been called Basilisk by the Boys, conjuring up images of an ill-disposed reptile, became her close friend. The bond between them strengthened because they shared a common recent tragedy. Her father and his mother had each been taken by SS troops in Rome the previous October and sent to concentration camps in Germany. Because of the wartime embargo on mail, they had learned the news only a month earlier, in July. Mail had become a virtual lifeline now that three years of wartime silence had been broken. Finally they were able to receive letters from Italy and learn about the respective fates of their parents. Their spirits had been buoyed by the fall of Fascism the previous September, but their wishes for a brighter future in Italy were quickly dashed.

The Allies' invasion of Sicily, ending victoriously in August 1943, led the Italians to conclude that the war was lost. The Fascist Grand Council had deposed Mussolini on July 24. Secret negotiations for an armistice were set in motion and concluded on the eighth of September. Announced publicly, this was the event that had inspired such hope in Laura and Emilio. But the result of the armistice was that the Germans, feeling betrayed, had acted quickly to take charge of as much of Italy as they could.

The country was effectively placed under German military command, an occupation marked by severe reprisals and massacres of civilians. With fierce fighting between the Germans and the Allies taking place in the fall and winter and heavy losses suffered by both sides, the Allies' invasion was stymied. Rome and every part of Italy north of the city remained under German control.

The situation in Italy had initially not seemed menacing, but many Jews had already gone into hiding, either by procuring false documents or by finding a safe refuge. Laura's sister Anna and her husband, Alberto Montel, fled with their two young children from Torino to Switzerland. After living in comfort in Italy, they fell on hard times: they were short of money. Laura and Enrico began to send them a monthly remittance of $100, quite a significant amount of money then.

In Rome, the state of affairs for Jews was rapidly deteriorating. The first sign of alarm came when the Germans seized the registration records of Jewish members from the city's main synagogue. And then they struck. On the sixteenth of October 1943, at 5:30 in the morning, SS troops surrounded the old ghetto, where many of the poorer Jews still lived, and began going house to house arresting those they found. Loading them onto transport trucks, they brought them to the court-yard of the nearby Military College. The SS troops then set out to arrest the Jews living elsewhere in the city. A little more than a thousand Jews, some ten percent of those living in Rome, were swept up in the roundup on that October day.

Emilio Segrè's parents had been warned of the house arrests, and while his father waited in their escape car, his wife—hurrying back to their home to collect some jewelry—was intercepted by SS troops and taken away. His father, who had witnessed the arrest, could do noth-ing. He was driven away by their chauffeur, taken to a monastery, and remained hidden until his death a year later.

The SS had also arrested Laura's father, Admiral Capon, in the roundup. As the situation for Jews began to deteriorate, Laura's three siblings and their families had already gone into hiding. Fermi's sister, Maria, repeatedly offered to hide Laura's father. Partially paralyzed after a stroke and still believing his rank in the Italian navy would pro-tect him, the admiral had declined.

It took the Allies four attempts before they could finally break through the German lines. On the fifth of June 1944, they entered Rome, greeted there with jubilation by its citizens. And communication with the United States was finally restored. That is when the Fermis heard, both from Amaldi and from Fermi's sister, Maria, about the arrest of Admiral Capon. Some embers of hope continued to burn, as in a November 15, 1944, letter from Laura's brother-in-law that says the admiral "sent a card from Ferrara and we think he was sent to Theresienstadt. We have not had any news since then. We can still hope he'll be exchanged for German prisoners by the Allies. A request has been presented to the Jüdische Flüchtlingshilfe [Jewish refugee agency] and it might be good for you to try and get some information from competent sources in Washington."

Fermi's first letter to Edoardo Amaldi after the liberation of Rome comments on Laura's state of anxiety: "As you can imagine, Lalla [Laura's nickname] has been very pained by the news of her father; the uncertainty of his fate has been much worse than knowing him dead." The letter's phrasing "Lalla has been very pained" rather than "we have been very pained" is a telling reminder of Fermi's reserve in emotional matters. Equally striking is that Laura, in her engaging book, makes no mention of this.

The uncertainty about exactly what had happened to the deported Roman Jews in October 1943 would hover until the end of the war. Their ultimate destiny was feared, but not substantiated until a few survivors gave testimony about their gruesome deaths. Laura's father was murdered in an Auschwitz gas chamber immediately after arriving at the camp. And Emilio's mother never made it to Auschwitz, perishing in a rail car on the way there.

Laura and Emilio hardly spoke, not even to family members, about the deaths of their respective father and mother amid the horrors of the Holocaust. But when Laura and Emilio reconnected years after their youthful days in Rome and found themselves on top of an isolated mesa in the great American Southwest during the summer of 1944, they commiserated—and waited nervously for letters from Italy with hopes of positive news.

30

GÖTTERDÄMMERUNG

When Richard Wagner's turbulent opera *Götterdämmerung* premiered in 1876, German audiences were undoubtedly shaken by the doomsday scenario of the last part of his famous *Ring* cycle. The world was engulfed in flames, its ultimate destruction sealed. It was a nightmarish scenario, and the Los Alamos scientists could readily imagine Hitler cast as the pivotal evil force. Fears that Germany would develop the bomb before the Allies succeeded in doing so were particularly pervasive among Los Alamos's refugee physicists. Two factors came to bear: the first, a definite unknown, was how close the Germans were to having a bomb and the second, more apparent, was the many obstacles Allied scientists needed to overcome before they could succeed.

The physicists on the mesa would have been shocked to learn the true state of affairs of nuclear research in 1944 Germany. It was nowhere close to developing a bomb. The German workforce of outstanding physicists, many of whom were Jewish or married to someone Jewish, had been seriously depleted by racial laws. Organizationally, there was nothing that approached the scope of the Manhattan Project. The Uranverein (Uranium Club), formed in 1939 to explore fission, had

morphed into various structures that attempted to develop and produce nuclear weapons. However, they had never even achieved the
basic precursor: a sustained critical nuclear reaction such as Fermi
and his colleagues had accomplished with the pile in December 1942.

In regard to steps toward for making a bomb, the Germans failed to
appreciate, as Szilard and Fermi had, that carbon obtained commer-
cially was likely to contain hidden impurities that needed to be removed
before it could be used as a moderator in a pile. In addition, the Allies
were partially successful in preventing the Germans from having access
to heavy water, the alternative moderator.

The leadership of the German atom bomb project was also wanting. In June 1942 it was essentially entrusted to Werner Heisenberg,
a theorist who had little understanding of experimental trials. Paul
Harteck, one of Germany's leading experimenters working on the
German bomb project, put it succinctly: "How can you be a leader in
such technological matters when you have never run an experiment in
your whole life?" In contrast, Oppenheimer—also a theorist—adeptly
ran the Los Alamos project, aware of his limitations and ready to resolve
them. Heisenberg had no such comparable qualities.

Nor did Heisenberg possess Fermi's facility for rapidly estimating
the order of magnitude of any physical phenomenon. Rudolf Peierls,
who knew Heisenberg well from having worked as a student with him,
observed, "Though a brilliant theoretician he was very casual about
numbers." This was a disastrous flaw in a project that depended so
much on exactitude. Because of its own failed attempts, the Uranverein was confident that the Allies would not be able to develop the bomb.
On the other hand, the Manhattan Project was very much driven by
the belief that the German enemy could beat them to the bomb.

The vast enterprise of the Manhattan Project had been created with
a mix of major fiscal and human resources, a large dose of optimism,
and the commitment of many of the world's top scientists. In June
1944, counting construction workers, machine operators, military personnel, and scientists, the project had altogether 129,000 employees, a
figure dwarfing the effort the Germans mounted. Among the scientists
in the Manhattan Project were leading physicists from England and
Canada, known collectively as the British Mission, who had joined

with their American counterparts in a spirit of cooperation. In Los Alamos, the British team was led by James Chadwick, the discoverer of the neutron, and it included Frisch and Peierls, the two physicists who had alerted their peers that building a fission bomb was within reach.

Nevertheless, Allied bomb development was in jeopardy during the summer of 1944. Those at Los Alamos were plagued by doubts of reaching their goal of having several bombs ready by the middle of 1945, the time estimated for them to still have an impact on the war. The strategy that had been developed was to proceed on two fronts, one using U-235 and the other plutonium, but it was beginning to look as if neither would be successful in time.

Difficulties had been encountered both in the production of sufficient fissile material for the bombs and in the development of a bomb detonation mechanism, two interrelated aspects of the problem. The separation of U-235 from natural uranium at Oak Ridge was proceeding so slowly that by mid-1945 there might be enough for only one uranium bomb, with no prospect of having more soon after. Production of plutonium was more promising, but the Hanford reactor was not scheduled to enter into operation until the end of September. If it did not function properly, there would not be enough plutonium available for even one bomb, much less for several of them.

Detonation mechanisms, the other major cause of concern, had already been discussed at the Los Alamos orientation meeting in April 1943. Two mechanisms had been seriously considered. The first, the so-called gun mechanism, was relatively straightforward. A piece of fissile material with a Fermi neutron reproduction factor k less than one would be shot at a second similar piece placed near the first. When united, the two subcritical pieces would become supercritical, with k closer to two—that is, two neutrons would be produced for every one that was absorbed. An explosion would occur at an exponential rate, in microseconds. The bomb needed to be small enough to be carried by a plane.

The second mechanism hinged on being able to surround with explosives a sphere of fissile material not dense enough to be critical. When simultaneously ignited, the explosives would cause the sphere to implode and quickly reach the density at which it became supercritical. But placing them so as to have uniform compression of the fissile material was extremely delicate. The implosion method was largely ignored in 1943 because it was intrinsically much more complicated than the gun mechanism and there was uncertainty as to whether it could be developed. In any case, it was felt at the time that the gun mechanism would work for plutonium as well as for U-235.

The detonation crisis in the summer of 1944 was precipitated by the realization that though the gun mechanism would function as anticipated for U-235, it would not do so for plutonium. The plutonium produced in reactors contained too large a percentage of the radioactive isotope Pu-240. It would vitiate the gun mechanism by prematurely detonating; the bomb would fizzle. There was also no hope of separating the desired isotope of plutonium, Pu-239, from Pu-240.

After more than two years of work by many brilliant scientists, an unprecedented mobilization of resources, and hundreds of millions of dollars having been spent, there was a grave threat that the Manhattan Project would have to be declared a failure on a monumental scale.

The circumstances were so dire that in mid-July 1944, Oppenheimer notified Conant, the head of the National Defense Research Council, and arranged a meeting in Chicago with him, Compton, Fermi, and Groves to discuss how they might salvage the project. Their conclusion was that Los Alamos needed to undertake a crash program to develop another detonation mechanism for the plutonium. There was only one real candidate: the implosion mechanism. If that could not be made to work, there was no substitute.

Oppenheimer had hoped to have Fermi come to Los Alamos full-time in August 1944 to help the laboratory overcome the crisis, but Fermi's presence in Hanford had to be the Manhattan Project's overall priority. Since its reactors were scaled-up versions of CP-1 and CP-2, he had been a vital guide for the building of Hanford's reactors. If they did not operate properly, there was no backup. The whole project was poised on the edge of a ridge, with chasms on both sides. There would be success or utter failure. The Hanford team would understandably have been even more concerned had Fermi not been present in September when they would try for the first time to have the reactors reach criticality. As Compton asserted, "Enrico Fermi was our anchor man on such occasions."

The reactor did go critical as scheduled, a few minutes after midnight on Wednesday, September 28. Everything was working as expected. But after operating for a few hours at a higher level than any previous reactor, trouble set in: the power output was inexplicably decreasing. The operators started moving the control rods out to keep the reactor steady, but that helped for only a little while. By morning the reactor was completely dead.

A distressed Crawford Greenewalt, directing the Hanford Project for DuPont, turned to a friend driving with him to the site the next morning and said that it could not be due to any problem with the properties of the material they were using "since this had clearly been well worked through by Fermi." It had to be something else: water leaking into the reactor from one of the pipes, or perhaps something wrong with the water itself. The growing suspicion that the stopping was due instead to the material was confirmed in everyone's mind early Thursday morning when, seemingly miraculously, the pile began operating again. It went critical at 7:00 a.m., dying once again twelve hours later.

The Princeton physicist John Wheeler was Hanford's local nuclear physics expert in 1944. He had considered that this might occur but had thought the possibility was remote and had not found a way to pretest the reactor. When it did occur, Fermi and Wheeler agreed on what must be taking place: the reactor was being poisoned.

The sequence had to be that after absorbing a neutron, some of the U-238 nuclei in the reactor turned into U-239, which then decayed in two steps to plutonium 239, the desired end product. However, other U-238 nuclei underwent fission. In one of the decay chains that followed, a formidable absorber of neutrons was being produced, so powerful that it could absorb all the neutrons being produced, effectively shutting down the reactor. But then why did the reactor turn on again after a few hours? There was an answer for this as well: the absorber was unstable. Within a few hours, its nuclei decayed into other nuclei that did not absorb neutrons. The reactor would then start up and the cycle would repeat itself.

Fortunately there was a solution. Eugene Wigner had designed the reactors for maximum efficiency, but on Wheeler's advice, DuPont had planned on having a margin of safety by installing additional material and equipment should they be needed. Time would be required to make the changes and connect the necessary extra water for cooling, but it could be done reasonably expeditiously. And this would fix the reactor because, as Fermi and Wheeler now calculated, making the changes would allow it to overcome the poisoning.

In early October, after Fermi felt assured the process would work and had convinced his colleagues this was so, he left for Los Alamos. He was now ready to work on helping the laboratory develop the detonation mechanism.

Fermi's arrival in Los Alamos was heralded by a half dozen scientific luminaries meeting for lunch one day. Edward Teller had announced, "It is quite certain now that Enrico will arrive next week." Stan Ulam, an outstanding mathematician who had left Poland, had heard Fermi referred to as the Pope and immediately intoned in perfect Latin the classic announcement of the election of a new Pope made by the Vatican's senior cardinal from the balcony overlooking St. Peter's Square. Once the group of scientists understood the reference, they broke into applause.

When Fermi arrived on the Hill, Oppenheimer appointed him the laboratory's associate director and also created Division F, where the F stood for Fermi, the universal consultant. He was probably the only person with expertise in every aspect of the physics problems being faced in the laboratory. Preparation of samples, theory, electronics, computing, optics, chemistry, and hydrodynamics were all in his repertoire. And of course Fermi was also known to be a marvelously clear and patient expositor.

Though acting as an informal adviser to every division, Fermi was not a significant contributor to developing the implosion mechanism. But he helped in many other ways. One was by assisting Oppenheimer in resolving administrative difficulties that arose from the lab's reorganization. At Oppenheimer's request, Fermi had met with a leading scientist to try to convince him to move into a new administrative role. As a lure, Fermi agreed to meet him every Friday after lunch to discuss physics. At that point, the reluctant scientist confessed, "I was ready to sell my soul." Bob Wilson subsequently became an excellent administrator.

Meanwhile, Edward Teller was a frustrating obstacle for the Theory Division's head, Hans Bethe, by having, as Bethe said with displeasure, "declined to take charge of the group which would perform the detailed calculations on the implosion." Calculations needed to be done, with or without Teller, for the mission to have a chance of success. Oppenheimer handled the volatile situation with the touchy Teller by appointing the proficient and cooperative Peierls to replace him. At the same time, Oppenheimer urged Teller to stay on at the laboratory rather than resigning, as he was threatening to do. With Fermi in agreement, Teller was appointed as head of a new group in Division F. One by one, every thorny situation was being handled.

As predicted by Fermi, the Hanford reactor began producing plutonium in November 1944. By early 1945, confidence was growing that the implosion mechanism would work. It had taken an organizational overhaul of the laboratory and the dedicated labor of many extraordinarily talented individuals; enormous impediments had been overcome. But still, the laboratory felt a definitive test of the plutonium bomb was key. Although draining resources and consuming a large fraction of the lab's precious plutonium, it would quell nagging doubts. A test was scheduled for mid-July, only weeks before the bomb would be used against the enemy, assuming all went well.

The specter of Hitler reigning over the destruction of the world was fading. If there were to be a Götterdämmerung, at least it would not be in the hands of a maniac. It would be in the hands of a novice president from America's Midwest, a man known for his modesty and honesty. Harry Truman had only experienced the vastness of the Southwest from the inside of a train car and regarded it as "mighty pretty country." On July 16, less than three months after he took office, a blinding flash lit the New Mexico desert at 5:29 in the early morning. The "Twilight of the Gods," as Wagner's opera is translated, had been introduced by Los Alamos's best and brightest.

31

THE HILL

When Fermi came to settle in Los Alamos in October 1944, he traveled under his usual guise as Mr. Farmer and was accompanied by his trusted bodyguard, John Baudino. Only needed when Fermi went outside the confines of the secret city, Baudino lived with his wife and baby daughter in Los Alamos while he worked at the security office there.

There was little time for Fermi to enjoy being reunited with Laura and the children. He was instantly immersed in lab work, true to his reputation among colleagues as a man "totally absorbed, taking little notice of his family." This observation was shared by Laura, who wrote with no obvious resentment, "It was typical of Enrico to be engrossed in his work and to pay no attention to what was going on around him."

Fermi was aware, and relieved, that Laura had adapted well to life on what was affectionately called the Hill. She had connected, or rather reconnected, with families whose friendships dated back from days in Rome such as the Segrès, Bethes, Peierlses, Rossis, and Tellers—all Jewish refugees.

The Fermi children had settled into school, Nella somewhat miffed because two different high school grades had been combined into the same classroom. Inevitably, as is almost always the case with children, Nella found her way to mischief. The high wire fence surrounding the town was a magnet for daring; childhood pranks included sneaking out and venturing briefly into surrounding canyons. It was Los Alamos kids, not spies or saboteurs, who knew where the holes in the enclosure were. In playing games with other peers, Giulio decided it was time to adopt a more American name; from now on, he declared he should be called Judd.

Although Laura had warned him, Fermi began to appreciate over time that life in Los Alamos was on the primitive side. Accommodations were minimal, electricity was sometimes iffy, the houses had no telephones, there were frequent water shortages, and the mud in the streets was often so thick that cars got stuck. A newcomer shared her first impressions: "The rickety houses looked like the tenements of a metropolitan slum area; washing hung everywhere and the garbage cans were overflowing." But there were compensations. Laura wrote to a friend, "Through the three contiguous windows of our living room I could see the round green tops of the Jemez hills slanting against the sky, as in a three-panel picture by an old master."

The wife of one of the scientists wrote an amusing essay entitled "Not Quite Eden"; as she noted, "We managed to adjust ourselves to the oddest conditions under which a community has ever been maintained and within these limits to lead reasonably normal, happy lives." Sirens marked the beginning and end of the workday and signaled a lunchtime break from noon to one. Guards at the main gate, controlling access to the community, scrupulously checked resident passes. Except for official business, traveling much farther away than Santa Fe was not permitted. Nor were outsiders allowed to enter, other than those employees who commuted there or a group, such as cleaning personnel from nearby pueblos. There was no mailman or milkman, no unexpected knocks at the door from traveling salesmen.

The Hill was a unique combination of democracy and hierarchy. All of its citizens shared a common mission, had experienced the upheaval of moving, and weathered the town's hardships. Nonetheless it was a stratified society: "Lines were drawn principally not on wealth, family or even age, but on the position one's husband held in the

Laboratory." There was a distinct pecking order throughout the war years that continued to some extent after the war. Physicists—especially theoretical ones—were on top, followed by chemists, technicians, and computing specialists. And the split between military and scientific personnel was pervasive, with a few exceptions. The divisions were ethnic as well. As a teenager, Nella Fermi took note that maintenance personnel were mainly Spanish Americans and that they lived in largely segregated neighborhoods in the Atomic City.

A notable exception cutting through these barriers was regular community square dances. They were a fun pastime for all backgrounds and ages. Fermi particularly came to enjoy the dances, as they suited his dual predilections for informality and for being American. It was also an occasion the elder Fermis attended with Nella, who typically brought her Spanish American best friend. Fermi's introduction to square dancing was somewhat inauspicious, though characteristic of him. He sat on the sidelines until he was sure he completely understood the calls, the patterns, and the sequences. He then asked one of the experts to dance with him. She recounted that he had successfully mastered the steps but "danced with his brain instead of his feet." The feet would come later.

The seclusion of the town led to an exuberant social life on the Hill. There was little to do for entertainment. In the town's theater, movies could be viewed from hard seats for fifteen cents; the theater doubled as a gymnasium, dance hall, or overall performance center. Parties of every size and shape became the amusement of choice. As one wife noted, "Saturday nights the mesa rocked with a number of dances and parties." Since residents could not talk about work, they could let off steam from the strain of the heavily cloaked project. The parties were often raucous, with liquor flowing liberally and spirits high.

Los Alamos was a community composed largely of young couples. The average age of the scientific employees was twenty-nine, and almost none were over forty. Many of them started families. During the first year of its existence, reportedly eighty babies were born on the Hill. Their birth certificates read that they were born at Post Office Box 1663, Santa Fe. It is not altogether surprising that years later Nella Fermi wrote her dissertation on patterns of fertility. The imprint of Los Alamos asserted itself in unexpected ways.

With the population of Los Alamos doubling every ninety days, it was almost impossible to provide adequate housing and services for residents whose numbers had reached 5,700 by the end of the war. The buzz was that Groves had even ordered Oppenheimer to see to it that fewer babies were born in order not to overwhelm the limited medical resources available. The story was made more piquant because not only had Oppenheimer refused to carry out the directive, but he and his wife, Kitty, had a baby girl, born on December 7, 1944.

Rumors about what was happening on the mesa flew. Santa Fe, a mix of Anglo, Native American, and Hispanic ethnicities, added a new breed to its multiculturalism: "Hillers." Although Hillers tried to blend in, they invariably stood out on their excursions to the delightful city of light, whether because of their European accents, city dress, or general quirkiness. Santa Fe gossip about Los Alamos included speculation that it was a home for pregnant WACs (Women's Army Corps) or, more facetiously, a building site for a submarine. As a ploy, Oppenheimer encouraged the Hillers to talk loudly about preparing an electric rocket. Noise from explosions made the latter somewhat more plausible.

The countryside surrounding the Hill was magnificent. There was fabulous hiking, abundant fishing in trout-filled streams, horseback riding opportunities, and mushroom hunting in the pine forests. Winter ice skating on a frozen pond was at Fermi's doorstep. A roughly shorn ski slope, rigged with a simple rope tow, had been forged by physicists with the help of GIs.

Fermi was happy at Los Alamos because he felt in his element, both because of the physical surroundings and the intellectual stimulation. It felt youthful and pioneering, mirroring the days of Dolomite excursions and the bonds formed with the Via Panisperna Boys. Emilio Segrè, one of the Boys and an avid fisherman, tried to convince his fellow emigrant of the pleasures of this sport. It would be an excellent way for the two of them to relax over the weekend after intense days of working at the lab. Fermi, after a few attempts, displayed neither any interest in fishing nor any luck with it. Segrè, waxing eloquent about the intellectual challenges of fishing, proclaimed, "You see, Enrico, it's not so simple. The fish are not stupid, they know how to hide. One has to learn their tricks." Fermi grinned. "I see, matching wits!" Clearly, Fermi's wits did not match the fish's; he never managed to catch one.

Fermi's preferences were hiking in the summer and skiing in the winter. While wives and children sometimes joined weekend outings, the company was most often all male. Otto Frisch, part of the British team working at Los Alamos, described what it was like being with Fermi on weekends when "he was usually out walking with a group of young people who felt entirely at ease with him, though he was obviously the master. I have never met anyone who in such a relaxed and unpretentious way could be so dominant."

During the rest of the week Fermi played the role of internal consultant to the lab and "never appeared to be in a hurry, yet he got a great deal done because he was so organized." His routine would be to be available mornings in his office for questions and work afternoons in the laboratory, wearing his lab coat to signal his own research priorities.

Fermi enjoyed sparring with young physicists, with a smile on his face and his eyes darting quickly about. This is illustrated by a story the celebrated physicist Richard Feynman told of an early meeting with him. Feynman's braininess was promptly recognized at Los Alamos, and at age twenty-six he became a group leader, but in 1943 he was still a new laboratory recruit who had just finished his Ph.D. One day, while he was having difficulties understanding some results he had obtained, Fermi stopped by.

Feynman described what followed: "I told Fermi I was doing this problem, and I started to describe the results. He said 'Wait, before you tell me the result, let me think. It's going to come out like this (he was right), and it's going to come out like this because of so and so. And there's a perfectly obvious explanation for this'—He was doing what I was supposed to be good at, ten times better. That was quite a lesson for me."

Fermi had obviously not resisted the chance to show off a little, but someone of Feynman's intelligence was not threatened. A symbiotic friendship between the two men, with similar views about informality and problem solving, flourished in Los Alamos and would continue afterward.

After ten years of intense work on neutron physics, Fermi was thrilled to be exposed to wide-ranging ideas. He was freer to explore fields related to other aspects of bomb research. One that particularly caught his fancy was the possibility of using new tools to perform the complicated calculations necessary to achieve implosion.

Richard Feynman and John von Neumann, known respectively as Dick and Johnny, shared a fascination with machine computations, an embryonic forerunner to what became the computer revolution. They quickly folded Fermi into their web of curiosity. Johnny knew of exciting research being conducted at Bell Laboratories, Harvard, and the University of Pennsylvania. An expert in explosives and hydrodynamics, the rotund and immaculately dressed Hungarian moved easily and with alacrity from field to field. This latter trait was shared by Dick and Enrico, and the three geniuses pursued the complexities of mechanical mathematics with mutual humor and virtuosity.

While Fermi naturally gravitated toward his young colleagues, he also was pleased to see Niels Bohr make prolonged visits to Los Alamos. Bohr had escaped Nazi-occupied Denmark after being warned that the Third Reich considered his mother, and accordingly him, to be Jewish. A towering figure in international physics, the almost-sixty-year-old had been left a man without a country, traveling extensively in America and England. Under the alias of Mr. Nicholas Baker on his Los Alamos visits, Bohr relished the natural beauty and quietude of the New Mexico landscape, enjoyed hikes and skiing and especially the opportunity to speak physics on a one-to-one basis with the Los Alamos scientists.

Bohr was treated as a wise elder, particularly by those who had been under his tutelage in Copenhagen. At a time when morale in Los Alamos was low, Bohr bolstered spirits by urging them onward. As the eminent physicist Victor Weisskopf wrote, "Bohr immediately involved us in private discussions of the significance of what we were doing . . . his idealism, foresight and hope for peace helped us see sense in all these terrible things. He inspired many of us engaged in the work of war to think about the future and to prepare our minds for the task of peace that lay ahead." Fermi, always the rationalist and wary of inspirational messages, was less taken by Bohr's vision, commenting once to Ulam that when "Bohr talked he sometimes gave the impression of a Catholic priest celebrating mass."

The retrospective lens of history highlights extraordinary aspects of the Hill. Many are admirable, several of them controversial. Whether the scientists should have focused more on the moral implications of their work will be debated in perpetuity. But there is one lens that is absolutely clear. Los Alamos was a male-dominated society, reflecting the norms of the times and certainly characteristic of the physics world. Physics came first even on weekends. Domestic life was a distant second. The lives of the women resembled those of war wives; arguably, their husbands were more physically present, but not emotionally so, a situation exacerbated by the oppressive air of secrecy.

In historical accounts of the project, women tend to be barely acknowledged. As only one example, the computational functions so necessary to the project's success were staffed overwhelmingly by women. Their job description was listed as "computers." Yet one rarely hears about them. A division leader was reported to have said that for the exacting and exhausting calculations, "we hire girls because they work better and they're cheaper."

The title of a compilation of essays written by the women of wartime Los Alamos speaks for itself: *Standing By and Making Do*. And in another book, a chapter called "The Cult of Masculinity" documents that the expectation of Los Alamos women was to be primarily supportive of their husbands' work.

Still another book casts a negative shadow on Los Alamos wives, describing them as becoming "depressed, quarrelsome and gossipy" while at the same time acknowledging their loneliness and sense of omission from the secret work of their husbands. Purportedly Oppenheimer was concerned enough to consult a psychiatrist, who advised him to "keep the women busy and to pay them so that they would have tangible proof of their usefulness."

Whether jobs were viewed as therapeutic release or a way for women to contribute to the overall war effort, the wives of Los Alamos scientists were encouraged to become teachers, librarians, medical technicians, clerks and—as already noted—computers. Even Laura, the privileged child of well-to-do Italians, took a position. Never employed before, she worked part-time in the Los Alamos hospital for the doctor in charge of health for the Technical Area, the ultra-guarded laboratory area that required a special white badge for entrance.

When the lab's first radiation accident occurred there in August 1945, the young physicist Haroutune (Harry) Daghlian was rushed to the hospital unit where Laura was assigned. After an excruciating twenty-five days, the twenty-four-year-old died. Laura learned firsthand about the horrors of radiation poisoning.

Enrico had never talked to Laura about such dangers in the workplace. Nor did she have any idea that these dangers would spread well beyond the boundaries of the Hill. Tragically, in the same month and year that the young Armenian American suffered from lethal radiation poisoning in New Mexico, tens of thousands of Japanese experienced similar fates.

32

"NO ACCEPTABLE ALTERNATIVE"

The secrecy that enveloped the Manhattan Project seemed impenetrable. Billboards depicting the mythical monkeys that neither saw, heard, nor did evil, cautioned employees: "What you see here, what you do here, what you hear here, when you leave here, let it stay here!" This did not seem to deter one man, Klaus Fuchs, who as part of the British Mission had been recruited in August 1944 by the Los Alamos lab. He worked on the implosion of the Gadget. A valued member of the scientific team, he was considered polite and refined. Often he joined the Fermis and others for recreational and social gatherings. In the words of Laura, "we all trusted him and saw him frequently." All the while, he was passing classified information about the bomb to the Russians.

According to America's popular *Life* magazine, prior to August 7, 1945, "no more than a few dozen men in the entire country knew the full meaning of the Manhattan Project, and perhaps only a thousand others even were aware that work on atoms was involved." Whether *Life* considered Fuchs as one of the few dozen men who knew is unspecified, but it would have been correct to count America's vice president among the clueless—that is, until April 12, 1945.

That was the day when President Roosevelt suffered a massive cerebral hemorrhage. He died a few hours later. Harry Truman, a former U.S. senator from Missouri and vice president for little over a year, was sworn in as president that evening.

In a cursory remark made after the swearing-in, Secretary of War Henry Stimson informed the newly minted president of "the development of new explosive of almost unbelievable destructive power." Perplexed and startled, Truman had known nothing about it. As chairman of the Senate's Committee to Investigate the National Defense Program, he had been aware of the Manhattan Project's existence but had not been told what it was trying to achieve—nor about its progress.

After detailed briefings, Truman added the bomb to an already full agenda. The war in Germany was ending, but the one with Japan was not. The relationship between the United States and the Soviet Union had become increasingly tense as arguments about the organization of postwar Europe rose to the surface. Given these circumstances, there was little sympathy in high government circles for Bohr's advocacy of international cooperation about nuclear weapons research.

The world picture shifted significantly on May 7, 1945, when Germany surrendered. Hitler had committed suicide the week before. The fiendish dictator was dead and Germany had not developed an atomic bomb, a fear motivating Los Alamos's refugee scientists. The raison d'être for their relentless drive and herculean efforts had suddenly vanished. They were stunned, both elated and deflated: delighted the war in Europe was over and confused about further justification for their atom bomb work.

Yet no one quit the project in May, or in the months that followed. At the end of 1944, Józef Rotblat, a Polish physicist with British citizenship, had done so. He left the lab on the basis of it becoming evident that Germany had abandoned its bomb project and "the whole purpose of my being in Los Alamos ceased to be." Contemplating why others had not departed, he attributed it to a wide range of reasons, from sheer scientific curiosity about whether the bomb would work to the argument that its use would save American lives.

But Los Alamos was a project under the aegis of the military, and the war against a stubborn enemy was not over. Martial strategies and discussions dominated the dialogue. Washington hoped this new weapon would bring the war to a swift conclusion. On the twenty-third of April, Groves, with his usual gruff manner, had updated his patrician boss, Stimson, on activities at the laboratories. He predicted that by early August, enough U-235 would have been produced at Oak Ridge for one bomb. Hanford's plutonium output was proceeding at a rapid rate. Difficulties at Los Alamos with the implosion mechanism were being overcome. There was no assurance that the plutonium bomb would work, but if a test showed that it did, an implosion bomb would also be ready by early August. The report was positive on all counts.

Two days later Stimson and Groves had a meeting with the president, briefing him thoroughly on the Manhattan Project and suggesting that a civilian committee be formed to provide recommendations to him about uses of the bomb, postwar research and development, international control, and release of information to the public. The president readily agreed.

The first meeting of the Interim Committee, known as such because it expected to disband after the war ended, was held at the Pentagon on May 9. With Stimson as chairman, committee members included two high government officials, the president of MIT, and the two men most responsible for formulating science policy during the war, Vannevar Bush and James Conant. James Byrnes, about to be named secretary of state, was placed on the committee as President Truman's direct representative. And though not a member, the Armed Forces chief of staff, General George Marshall, attended the meetings to gauge how its deliberations would affect the military.

It was a tall agenda, especially when the topic of the bomb surfaced. This elite group would make judgments reverberating throughout the world in both the moral and military realms. There was no question within government circles that the bomb should be used. *How* to do so was still debated.

The next meeting of the Interim Committee was a two-day session at the end of May, when they expected to have input from a specially constituted four-person Scientific Panel composed of J. Robert Oppenheimer, Enrico Fermi, Arthur Compton, and Ernest Lawrence.

It is notable that Fermi was the only one on the panel whose mark on the Manhattan Project was not linked to major administrative responsibilities and heavy-duty policymaking. At Los Alamos, he had been able to focus on what he liked best: physics, pure and simple. Being named to the panel, with its somber political agenda, was not what Fermi sought. He was among those scientists "not bothered by moral scruples" as described by the disillusioned Rotblat, "quite content to leave it to others to decide how their work would be used." Most probably, Fermi's appointment was made precisely because he represented a neutral, dispassionate approach.

In their May 31 meeting, the Interim Committee asked the Scientific Panel if the Japanese should be made aware that such a weapon existed. Or should the bomb be dropped on a Japanese city without warning, much as the Japanese had not warned about Pearl Harbor? Or would a demonstration on an uninhabited remote site be adequate to convince the Japanese of the futility of further fighting?

These same questions were posed the following day to a group of four prominent industrialists. In addition, the committee addressed general questions such as whether the Russians should be informed about the bomb, what type of atomic energy buildup would be appropriate in the postwar period, and how research should then be organized.

At the end of two days of meetings, Truman's personally appointed representative, Byrnes, made a beeline straight to the president, telling him the Interim Committee had agreed that the bomb be used on a city "without prior warning." In the eyes of this seasoned politician who had served thirty-two years in public life including a one-year stint as a Supreme Court justice, there was no other possible conclusion. How could one justify holding back in wartime a weapon that could prevent American lives from being lost? And how could one account for spending two billion dollars on a weapon and then not using it? And finally, how could one possibly obtain funding from Congress for continued nuclear research in a postwar world if there was nothing to show for what had already been achieved?

A more nuanced approach was offered by Secretary of War Stimson, considered by Truman "a man of great wisdom and foresight . . . as much concerned with the role of the atomic bomb in shaping of history as in its capacity to shorten this war." While acknowledging that firebombings in Dresden (February 1945) and Tokyo (March 1945) had each killed approximately a hundred thousand people, Stimson was concerned that even if the Japanese death toll were no higher, this new weapon might be regarded as an atrocity of a different order. It would be the work of a single bomb whose destructive aftermath was branded by a new way of dying, the slow agony of radiation poisoning. In addition, if the Russians were not notified beforehand of the bomb's existence, they might construe the action as saying that they were no longer considered U.S. allies.

Believing that Byrnes was acting too impulsively, Stimson asked the Scientific Panel to meet again in mid-June and provide him with a written set of recommendations on the initial use of the new weapon. He urged them to consult with others in the scientific community who played a role in the Manhattan Project, but were not as invested in making the actual bomb as the Los Alamos group.

During the May 31 meeting, Compton had shared with the Interim Committee the position advocated by physicist James Franck, his Chicago colleague. Franck, one of the many Jewish scientists who had fled Germany, was, as his grandson wrote, "sensitized to the ethical and political issues involved by his participation in World War I's chemical weapons program" and was strongly against military use of the bomb. His agreement with Compton to act as the Met Lab's chemistry division chair had been reached with the proviso that if a bomb was ready for use, Franck could express his opinion to someone at the highest policymaking level. Compton did not consider it an unreasonable request from the celebrated 1925 Nobel laureate. When he returned to Chicago from the Washington meetings, Compton asked Franck to head a six-man Met Lab Committee on Political and Social Problems that would write a report formalizing these ideas.

The Franck report, a hastily prepared document, argued for a demonstration showing the world the power of a nuclear bomb detonation. The test, in an uninhabited area, could stimulate a broad national and international public discussion about use of the bomb. The twelve-page statement was personally delivered by Franck and Compton to

Stimson's office in Washington on June 11, its summary stating, "We believe that these considerations make the use of nuclear bombs for an early attack against Japan inadvisable. If the United States would be the first to release this new means of indiscriminate destruction upon mankind, she would sacrifice public support throughout the world, precipitate the race of armaments, and prejudice the possibility of reaching an international agreement on the future control of such weapons." The report addressed head-on the probable political and ethical fallout from America dropping the bomb.

On June 15 and 16, the Scientific Panel met in Los Alamos. They had been asked by the Interim Committee to prepare a report on whether any kind of demonstration could be devised that would seem likely to bring an end to the war without using the bomb against a live target. The panel stuck to its mandate, although conversations lasting until the wee hours occasionally veered into the realm of politics and ethics. On these, Fermi remained largely silent, in keeping with his aversion to offering advice on such matters. However, on technical issues Fermi easily and profusely shared his acumen and insights.

The panel submitted its report posthaste to the Interim Committee, first presenting the different views about the atomic bomb held in their community, including those of the Franck report. Their conclusion contrasted sharply with the Franck report. Compton, Fermi, Lawrence, and Oppenheimer unanimously stated their finding that "we can propose no technical demonstration likely to bring an end to the war; we see no acceptable alternative to direct military use." Although a bit circumspectly presented, the panel's report helped pave the path to dropping an atomic bomb on Japan.

At the conclusion of the report, in a paragraph that echoed Fermi's belief that physicists lacked special insights into nontechnical matters, it equivocated:

> With regard to these general aspects of the use of atomic energy, it is clear that we, as scientific men, have no proprietary rights. It is true that we are among the few citizens who have had occasion to give thoughtful consideration to these problems during the past few years. We have, however, no claim to special competence in solving the politi-

cal, social, and military problems which are presented by the advent
of atomic power.

The panel's disclaimer of "no special competence" did not impede them
from recommending that there was "no acceptable alternative."

Whether the Interim Committee actually ever read the Franck
report is unclear. When Franck and Compton delivered it to Stimson,
he was out of town. Compton left a cover note expressing his own view
that the report had not weighed sufficiently the number of American
lives saved if the bomb was promptly used. Although not necessarily
meaning to undercut the report's contents, Compton's comment may
have effectively done so.

The moral aspects of dropping the bomb continue to be debated,
and are rendered more complex by whirling uncertainties of how close
the war against Japan was to a conclusion and the prospects of Russia's
invasion of Manchuria. But an enduring question is how such momentous decisions can be reached. They bridge military, scientific, political, and ethical domains. Fermi and Szilard, one a member of the
Scientific Panel and the other a signer of the Franck report, diverged
in this respect as they did in others. Szilard was adamant that scientists should be prime decision makers. Fermi, on the other hand, adhered
to the stance of "no special competence" and saw scientists not on top,
but on tap to other experts.

In the case of whether or not to drop a bomb on a civilian population, the decision-making framework spanned a little over two
months. The answer was affirmative; the die was cast. But it had not
yet been proved that the implosion mechanism would function. This
was crucial because although the scientists were sure the U-235
bomb would work, only one bomb of this type would be ready in the
next few months. Assuming that was so, the Japanese might deduce
it was the only nuclear bomb the U.S. Army had at its disposal and
continue to fight.

Knowing that they would need to test a plutonium bomb, Oppenheimer had already set in motion the search for an uninhabited desert
location to hold the trial. It had to be remote from populated areas for

safety reasons, also far enough away for security reasons, close enough for material and personnel from Los Alamos to have access, and large enough for the actual test. The site eventually chosen, measuring eighteen by twenty-four miles, lay in the New Mexico desert two hundred miles south of Los Alamos and sixty miles northwest of the city of Alamogordo. Oppenheimer, inspired by a John Donne poem, gave both the site and the eventual test the code name Trinity.

The first time they would have enough plutonium for the test was July; the sixteenth was the day eventually selected. On the seventeenth, Churchill, Stalin, and Truman would be meeting in Potsdam to discuss, among other matters, the fate of postwar Europe and the possible entry of Russia into the war against Japan. Knowing that they had this new weapon would allow the United States to take a tougher stance in their negotiations with the Russians.

Fermi was one of the few physicists who seemed relaxed as day dawned at Trinity on the sixteenth. Groves remembered being annoyed the night before by hearing Fermi say in his usual somewhat sarcastic way that "after all, it wouldn't make any difference whether the bomb went off or not because it still would have been a well-worthwhile scientific experiment. For if it did fail to go off, we would have proved that an atomic explosion was not possible." Groves concluded afterward that this was simply Fermi's way of easing tension in the camp.

The Gadget at Trinity functioned without a hitch. Fermi's calm calculation of the power of the detonation became legendary. The story also added to the perception that he was always in control of his emotions. At some level, he was able to treat the bomb blast as just another physics experiment.

Fermi later told Laura he had been concentrating so hard on this particular experiment that he had not even noticed the sound of the explosion. A mixture of relief, elation, and concern filled the hearts and minds of those who had witnessed the detonation.

Phase one of the experiment was over. After examining in the lead-lined tank what remained at Ground Zero and completing several other tasks, Fermi set out on the return trip to Los Alamos. He had to admit that even his legendary endurance had its limits. When he reached home that evening, he was "so sleepy he went to bed without a word." Laura had no inkling of what had happened. She commented on what he later told her about his return trip to Los Alamos, "For the first time in his life . . . he had felt that it was not safe for him to drive. It had seemed to him as if the car were jumping from curve to curve, skipping the straight stretches in between. He had asked a friend to drive, despite his strong aversion to being driven." Was this due to fatigue, delayed emotional reaction, or both? The answer may be that while Fermi was seemingly infallible, he was also human.

After the history-making explosion at Trinity, all efforts converged on preparing for the two types of atomic bombs, each one to be dropped on a Japanese city. Groves had already convened a Target Committee to decide where. He told its members to recommend no more than four prospective Japanese sites, with a "governing factor that the targets should be places which would most adversely affect the will of the Japanese people to continue the war." He also stipulated, "To enable us to assess accurately the effects of the bomb, the targets should not have been previously damaged by air raids." This ruled out Tokyo.

Of the four sites selected by the Target Committee, Groves favored Kyoto, but Stimson overruled him. This distinguished elder statesman, formerly secretary of war (1911–1913) and secretary of state (1929–1933), had also been governor general of the Philippines before assuming that post. During that time he had visited Kyoto and came to appreciate the significance of Japan's ancient capital to the nation. Stimson felt that, given the city's history, it should not be destroyed. Groves bowed to his insistence; Hiroshima was selected in its place as the number one target. The city had never been bombed, and it held an important army depot and port of embarkation in its industrial area.

Three weeks after the test at Trinity, early in the morning of the sixth of August, a B-29 bomber piloted by Lieutenant Colonel Paul Tibbets took off from the United States airbase on Tinian, an island in the Northern Marianas. On the previous day, the name *Enola Gay*, the given name of Tibbets's mother, had been painted on the fuselage below the cockpit, and Little Boy, a U-235 bomb, had been placed in the plane's bomb bay.

Little Boy was dropped at 8:15 a.m. on Hiroshima. The city's terrorized population viewed with shock, horror, and disbelief the spectacle the Los Alamos physicists had seen from a safe distance three weeks earlier, a bomb equivalent to twenty kilotons of TNT. An estimated one hundred thousand people out of the city's four hundred thousand population died as a result of the explosion.

Three days after that, on the ninth of August, Fat Man, a plutonium bomb, fell on Nagasaki with similar results. On the fifteenth, the voice of the emperor, previously unheard by the Japanese public, was broadcast on a disc recorded the day before. He announced Japan's surrender. An official ceremony was held aboard the battleship USS *Missouri* on September 2. World War II was over.

33

AFTERSHOCK

I saw a bright blast, and I saw yellow and silver and orange and all sorts of colors I can't explain. These colors came and attacked us and the ceiling beams of the wooden school along with the glass from the windowpane all shattered and blew away all at once.

—Michiko Kodoma, age seven, Hiroshima survivor

The general impression is one of deadness, the absolute essence of death in the sense of finality without hope of resurrection. . . . It is everywhere and nothing has escaped its touch. In most ruined cities you can bury the dead, clean up the rubble, rebuild the houses and have a living city again. One feels that it is not so here.

—Navy Captain William C. Bryson at Nagasaki, five weeks after the bomb drop

The first the general public learned about the discovery [nuclear fission] was the news of the destruction of Hiroshima by the atom bomb. A splendid achievement of science and technology had turned malign. Science became identified with death and destruction.

—Józef Rotblat, physicist formerly at Los Alamos

I don't believe a word of the whole thing.

—Werner Heisenberg, leading physicist of the German nuclear energy project

Among those most stunned to hear about the American atomic bomb were ten of Germany's leading physicists. Along with Heisenberg, they were incredulous when they heard the news. Presumably to prevent the Soviets from capturing them, the ten had been held by the British since early July at Farm Hall, a mansion near Cambridge. The British had placed secret microphones in the mansion's bedrooms and gathering places. All conversations were recorded, unbeknownst to the Germans.

Over dinner on the evening of August 6, the scientists struggled to decipher how the bomb could have been assembled and how they themselves had failed to solve the technicalities of producing one. These inquiries dominated the discussions, which were punctuated by continuing expressions of disbelief and condemnation. When one of the scientists opined that "it was dreadful of the Americans to have done it. I think it's madness on their part," Heisenberg had a quick rejoinder: "One can't say that. One could equally well say 'That's the quickest way of ending the war.'" Otto Hahn, whose discovery of fission made the bomb possible, was devastated that his prewar research had led to so much death and suffering. Hahn's précis that "I am thankful that we were not the first to drop the uranium bomb" was a sentiment seemingly shared by all but one of the ten.

The myriad worldwide reactions to the dropping of the bomb ranged from welcome relief that the war was ended to denouncements of the United States, accusing it of crimes against humanity. It was not just the number of dead, approximating the firebombings of Dresden and Tokyo, but the horrendous nature of their deaths from radiation. America, whose international reputation hinged on being a decent and fair nation, was thrust into the macabre role of a Frankenstein.

A 1945 Gallup poll conducted immediately after the bombing documented that 85 percent of Americans approved of using the new atomic weapon on Japanese cities. A 2015 survey found that 56 percent of Americans believe the use of nuclear weapons was justified. The issue continues to divide Americans, although fewer are now supportive of dropping the bomb. In the jarring light of hindsight, opposition and resentment have grown.

The prevailing view in 1945 America was expressed by Secretary of

War Stimson as "the least abhorrent choice" to reach the objective of ending the war quickly. Many accepted the argument that an invasion of Japan would have cost hundreds of thousands of American lives, but Stimson in a *Harper's Magazine* article seemed to inflate that estimate to even greater numbers, "over a million casualties to American forces alone."

Whatever the number, U.S. troops felt spared. Plans had been ripening for a major invasion of Japan, combining army, navy, and air force troops. A young lieutenant, Paul Fussell, who later became a noted historian, had been slated to be part of the assault and remembered, "When the [atomic] bombs dropped . . . we all cried with relief and joy. We were going to live. We were going to grow up to adulthood after all."

It was reasoned that the bomb had saved not only American lives, but also those of the Japanese who would have perished if fighting continued. And the hope remained that its horrific consequences would act as a deterrent to future wars. This viewpoint was touchingly expressed by a twenty-year-old Los Alamos technician in a letter to his mother:

> Well at last you know approximately what goes on up here. This new bomb may sound inhuman but . . . this thing will mean peace forever, even with the cost of several thousands of Japanese civilians' lives at the present. Let us pray that it will be unnecessary to use any more even on our enemy.

The secret city had been outed in the most dramatic way imaginable. In President Truman's August 6 statement announcing the bombing of Hiroshima, he divulged the magnitude of the Manhattan Project and the sites it comprised. For the first time, Laura understood the nature of Enrico's work, why he needed to have the bodyguard Baudino, and why the family had moved to Los Alamos.

Once peace was declared, Laura described the scene in the Atomic City as festive: "Children celebrated noisily, paraded through every single home, led by a band playing on pots and pans with lids and spoons." Joy about the new peace overlapped with pride about the bomb, the two intricately entwined. Wives, who like Laura had been uninformed, could now talk about what they and their husbands had sacrificed and achieved. Men who had kept silent suddenly could speak about their work.

Not everyone in Los Alamos felt celebratory. One scientist said, "It seemed rather ghoulish to celebrate the sudden death of a hundred thousand people, even if they were 'enemies.'" Others, particularly those who had witnessed the explosion at Trinity, could imagine all too well what it must have been like to be caught unaware at Ground Zero. An unhappy and nauseated Robert Wilson, who had earlier described coming to Los Alamos as "romantic," would not join the festivities. Feelings deepened after the news of Nagasaki's destruction, the second use of an atom bomb. Oppenheimer was reported as being a "nervous wreck."

With the extensive destruction in Hiroshima and Nagasaki, not only was the secret out but its results began to stir the social consciousness of the wives of Los Alamos. Laura recounted how the women sobered when "among the praising voices some arose that deprecated the bomb, and words like 'barbarism,' 'horror' and 'mass murder' were heard from several directions."

In probing her own emotions, Laura did not have easy responses to such qualms. Thirty years later, she pondered the ethical dilemmas inherent in dropping the bomb:

> But above all, there were the moral questions. I knew scientists had hoped that the bomb would not be possible, but there it was and it had already killed and destroyed so much. Was war or science to be blamed? Should the scientists have stopped the work once they realized that a bomb was feasible? Could they have stopped it? Would there always be war in the future? To these kinds of questions there is no simple answer.

One woman who had a decisive reaction was Fermi's older sister, Maria, who wrote a troubled letter to Enrico from abroad. After saying that everybody in Italy was talking about the recent events, she added, "All however are perplexed and appalled by its dreadful effects, and with time the bewilderment increases rather than decreases. For my part, I recommend you to God, Who alone can judge you morally."

Franco Rasetti, Fermi's best friend during university days at Pisa, expressed himself more harshly in a letter to Fermi's early childhood friend Enrico Persico. In April 1946 Rasetti wrote, "It seems almost impossible that people that I once regarded as endowed with a sense of human dignity have lent themselves to be the instrument of such monstrous degeneration. And yet this is so and they don't even seem aware of it . . . In my opinion these scientists, among them many friends of mine, including Fermi, might face severe judgment by history."

Rasetti never voiced this sentiment publicly, but one cannot help thinking it caused a schism between him and Fermi. Admittedly, their interests had diverged, but the two made little effort to see each other even after Rasetti moved in 1947 to a professorship at Johns Hopkins. Rasetti's mind-set was steadfast: science cannot allow itself to be co-opted for warfare, no matter how dire the circumstances. And Rasetti had been consistent. In 1943 the British physicists working in Canada on atomic bomb development asked him if he would join them. He declined and, as he wrote in that same 1946 letter to Persico, "There are few decisions I have taken in my life for which I have had less cause for regret."

Although knowledge about the bomb was in Rasetti's lexicon, it drew a blank for the vast majority of the American public prior to Hiroshima and Nagasaki. What was this new weapon, the atom bomb? Where had it been developed, who had built it, what was it made of, and were there other bombs like it?

With remarkable foresight, Groves had anticipated that Americans would want answers to these and other questions. He commissioned in 1944 a short history of the Manhattan Project that would also explain key scientific notions to the public. Composed in total secrecy and approved by scientists and censors, the report was eerily scheduled to be ready for release by August 1945. The lead author, Henry de Wolf Smyth, was a Princeton University physicist who had worked for the project and been an associate director of the Met Lab; Groves had assured him of complete access to Los Alamos, Oak Ridge, and Hanford. With President Truman's final approval, the report was made available on August 12, serendipitously three days after Fat Man was dropped on Nagasaki.

The Smyth report provided a background description of the laboratories, a summary of the basics of nuclear fission and chain reactions, and a brief history of what had transpired. To the great displeasure of scientists advocating open international exchanges now that the war was over, the government also erected barriers to information sharing. One of the reasons Groves had urged writing the report was his desire for controlling intelligence flow regarding the Manhattan Project. He made this abundantly clear in a foreword, underscoring that any scientific information beyond the report's contents would be in violation of "the needs of national security" and that anyone disclosing anything further without authorization would be "subject to severe penalties under the Espionage Act."

To everyone's astonishment, the Smyth report became a *New York Times* bestseller, remaining on the list for many months. It was read by many, including Laura, who was given a copy by Enrico, telling her it encompassed all he could say about the work of the past years. Laura, judging herself "stupid" for not having guessed, recalled when Emilio Segrè during a 1943 Chicago visit had caustically greeted her with "Don't be afraid of becoming a widow. If Enrico blows up, you'll blow up too." Other hints had been subtler.

The Manhattan Project scientists were astonished by the wide public interest in the project and the eagerness to understand what unleashing the power of the nucleus meant for the world. Again, Fermi was pushed to the fore, asked to become a spokesperson for the report with the other three members of the Scientific Panel. It was a role he neither relished nor sought.

In the meantime, not content to have the panel be the only voice heard in Washington, the Los Alamos scientific community began to mobilize. In spite of their isolation on the mesa and the secrecy of their mission, the scientists had not abstained from discussing among themselves the multitudinous ethical issues about the development and use of the Gadget. Perhaps it was Bob Wilson's Quaker background that had motivated him in late 1944 to propose holding internal meetings at Los Alamos dedicated to such discussions.

Although Oppenheimer had tried to talk him out of it, Wilson proceeded to put up notices in the lab focusing on "The Impact of the Gadget on Civilization." Since no official records were kept about these or other scientific colloquia that discussed the possible repercussions of the bomb, accounts were based on memory. At one gathering, according to Wilson, Oppenheimer in his usual soft and riveting voice argued that the war "should not end without the world knowing about this primordial new weapon. The worse outcome would be if the Gadget remained a military secret." At that time, Oppenheimer's outlook had seemed logical.

After Trinity, and more urgently, after the Japanese bombings, the younger scientists in particular wanted to be heard. On the thirtieth of August, approximately five hundred of them formed an organization called the Association of Los Alamos Scientists, known tellingly as ALAS. Within days they drafted a document emphasizing the dangers of an arms race and the need for international collaboration and cooperation, views they tackled during the Wilson meetings and views they had heard espoused by Bohr and supported by Oppenheimer. They asked Oppenheimer to forward their document to Secretary of War Stimson with an eye toward approval for publication. He did so on September 9, with an attached note indicating that though he had not participated in the writing, he was in agreement with its statements.

That was not the case with Fermi. According to Laura, he did not endorse many of ALAS's positions and accordingly had not joined them. For him, war was determined more by will than by weapons, by leadership rather than technical advances. He felt the times were not ripe for world government. ALAS members, regarding his decision not to join them as more a matter of temperament than of opposition to their views, did not hold it against him.

ALAS was shocked to hear at the end of September that their document had been classified. Oppenheimer reassured its members, telling them this action was prompted by the president's upcoming message on nuclear energy; out of courtesy, he should have the first word. On October 3, 1945, in an address to Congress, Truman promoted "directing and encouraging the use of atomic energy and all future science information toward peaceful and humanitarian ends," thereby setting the groundwork for future international collaboration.

Momentarily appeased, the members of ALAS again felt upended when later the same day the May-Johnson bill was proposed in Congress, establishing a nine-person commission to control atomic energy. While both civilians and military personnel were included on the commission, the bill was broadly regarded as a military power grab. Furthermore, the Los Alamos scientists were dismayed by the bill's penalties for security violations: ten years in prison and a $100,000 fine.

Oppenheimer nevertheless supported the bill and quickly convinced the others on the Scientific Panel to go along with him. Leona Woods Marshall suggests that Fermi did not need convincing. He had already decided that "any change is for the worse and that a change from the military to the civilian would come into this category."

With the war over, ALAS no longer wanted to be under the thumb of the military or to have their freedom to exchange information curtailed unless there were very, very good reasons. Anderson, Fermi's trusted lieutenant, voiced a typical reaction in an October 11 letter he wrote to an ALAS organizer; he thought the Scientific Panel's members "were duped when they urged the scientists to keep silent about the army's bill."

Wilson took the protest a step further by mailing the classified ALAS document directly to the *New York Times*; officially this was a security violation. The newspaper promptly published it on the front page. Wilson later wrote, "For me it was a declaration of independence from our leaders at Los Alamos, not that I did not continue to admire and cherish them. But the lesson we learned early on was that the Best and the Brightest, if in a position of power, were frequently constrained by other considerations and were not necessarily to be relied upon." The politically active Wilson was not charged with a security breach and was roundly praised for bringing the issue into the public domain.

Nobody was more consistent or more effective in opposing the May-Johnson bill than Leo Szilard. He organized, wrote, lobbied, and spoke repeatedly against military control of nuclear energy. By the end of 1945, his and others' efforts bore fruit. The president abandoned his support of the May-Johnson bill. Connecticut's Brien McMahon, the chairman of the newly formed Senate Special Committee on Atomic Energy, began drafting a countermeasure designed to put a civilian board in charge of nuclear energy.

Fermi had not become a member of ALAS, nor did he join its later offshoot, the Federation of American Scientists, an organization designed to broaden the ALAS membership base to include those who had worked at Hanford, Oak Ridge, and the Met Lab. As for Fermi's non-participation, Szilard commented, "The struggles of our times did not affect him very much, and he is no fighter." On a more charitable note, Fermi's stance was compatible with his extraordinary ability and pre-dilection to compartmentalize physics, devoid of any distractions— political or otherwise. His devotion to science was unwavering, and probing its complexities was inevitable. As Fermi observed, "What-ever Nature has in store for mankind, unpleasant as it may be, men must accept, for ignorance is never better than knowledge."

In an August 28, 1945, letter to his Italian colleague Amaldi, Fermi had portrayed his time at Los Alamos as "a labor of considerable sci-entific interest" and continued by saying that "having contributed to truncating a war threatening to go on for months or years has undoubt-edly provided a certain satisfaction." In the same letter he expressed his hopes that the bomb would have a positive impact on international relationships. Fermi's tone, understated and dispassionate, was more than familiar to those who knew him.

34

GOODBYE, MR. FARMER

It had been one of the most remarkable periods of our lives," according to Laura Fermi. But it was time to move back to Chicago; the mission had been accomplished. Regrets about leaving were mixed with looking forward to getting back to a more normal life. Laura remembers departing from Los Alamos on New Year's Eve, half an hour before the celebrated ball drop at Times Square ushered in 1946.

Suitcases packed, they brought with them Indian pottery and jewelry, paintings, weavings, and cacti, memory-laden souvenirs that would decorate their new Hyde Park home. After spending the night in Santa Fe and being driven to the small Lamy train station, they waited for the *Chief* to pull in. The squeal of its brakes broke the silence of the clear, cold morning. For Enrico, it felt odd to board without the presence of his bodyguard, Baudino, who had accompanied him any time he traveled outside the gates of Los Alamos. Mr. Farmer was no more. Approximately twenty-four hours later, the tired family of four disembarked in Chicago's busy Union Station, the skylight of its great hall arching high above marble floors.

They all knew the move would be an adjustment. Judd (formerly Giulio) had disliked Los Alamos, in part because it had meant yet another upheaval. There had been a plethora of moves for the youngster: first from Italy to New York City, then to New Jersey, Chicago, and Los Alamos, now once again back to Chicago. Nella seemed to adapt more readily than he to the many changes. Life again would be different for her and her brother. A certain freedom in what was arguably America's most secure and crime-free town would be replaced by the watchfulness of living in a city neighborhood where muggings were not uncommon. For Laura and Enrico it meant that dear friends would scatter and the esprit de corps that had characterized their lives would dissipate.

In a step that was both perceptive and brilliant, the University of Chicago had developed a scheme that would maintain the momentum of scientific wonder and craft a structure enticing the likes of Fermi. In mid-July 1945, the university's vice president, its dean of physical sciences, and Harold Urey, who was associated with the Met Lab's program of isotope separation, traveled from Chicago to New Mexico to meet with Fermi, his colleague Sam Allison, and Cyril Smith, who directed the lab's metallurgy group. Since the visitors couldn't get passes to enter Los Alamos, the six of them met for lunch in Santa Fe.

On a sunny terrace overlooking a landscape of mesas and mountains, the Chicago group proposed the creation of a new entity, the Institutes of Basic Research, comprised of three specific components: an Institute for Nuclear Studies, an Institute of Metallurgy, and an Institute of Radiobiology and Biophysics. Each institute would be devoted to a major area of research and each would be headed by a world-renowned figure, hopefully one who had played a prominent role in the Manhattan Project. The three scientists the university aimed to attract were, respectively, Fermi, Smith, and Urey.

Emulating the nature of the work at Los Alamos but without the restrictions of secrecy, it was a daring but promising plan for a university. Robert Hutchins, the school's president, was enthusiastic about the notion. Funding was not likely to be a problem because the climate was different from what it had been before World War II, and government backing for the physical sciences was almost certain to be forthcoming.

With the war ending, Fermi had been actively recruited by the University of Chicago and Columbia University. His already considerable fame was further boosted by what had transpired at Chicago and

Los Alamos. About to become the new department chair at Columbia, Isidor Rabi wanted to have him back at Columbia. Rabi regarded Fermi as the greatest physicist he had ever known, next to Einstein. Arthur Compton wanted Fermi for Chicago. About to leave his professorship there for a university chancellorship, Compton told Hutchins that "if Enrico Fermi could be persuaded to take over my professorship of physics, the University of Chicago would profit by the exchange."

The Chicago proposal, in process for several months, appealed to Fermi's penchant for cross-fertilization of ideas, the kind of collaboration and cooperation that marked Los Alamos. It would be a first, an interdisciplinary model crossing traditional departmental boundaries, connecting and integrating unique scientific perspectives. With its innovative underpinnings, how could it fail but to attract outstanding scientists and students? It was an offer Fermi could not refuse.

To the delight of the Chicago delegation, all three Los Alamos scientists agreed to the proposal, with Fermi placing the condition that his friend Sam Allison be named head of the Nuclear Studies Institute; he wanted to be freed from administrative duties. He was content with a professorship in physics. Teaching was his passion.

The formation of the Institutes occurred during a propitious period. The future of Los Alamos was uncertain and physicists both young and not-so-old were seeking new opportunities. They realized the time was opportune for change, especially since they had become associated with the success in building the bomb. Academia was appealing, especially if it was predicated on team-oriented research and benefited from generous funding.

The first of the recruits came from Los Alamos. Edward Teller, whom Fermi always found stimulating, was offered a professorship in the Institute for Nuclear Studies, as was Herbert Anderson. From Hanford, Leona and John Marshall came back to Chicago with research positions. From Oak Ridge, Urey packed up shop and made his way to the Windy City. Several of the young Los Alamos physicists enrolled in the Institutes' graduate programs. A collection of Los Alamos veterans very much to the Pope's liking would continue to work with him.

An appointment at the Institute for Nuclear Studies might have also been possible for Szilard, but Fermi let it be known that he was not in favor of such a move. Fermi wanted physicists who would work as hard and as steadily as he did and who also, like him, preferred to leave politics to the politicians.

At this point, Hutchins stepped in. An admirer of Szilard, he created a position for him, half-time in the Institute of Radiology and Biophysics and half-time as adviser to the newly created Office of the Inquiry into the Social Aspects of Atomic Energy. With a good salary and no teaching responsibilities, this was an ideal solution for Szilard.

In the six months after Fermi had accepted the Chicago offer, he stayed at Los Alamos winding up research and delivering a series of lectures to its remaining staff members. The talks spanned neutron physics from the beginning to the latest developments. The work done had many useful and game-changing applications to improve living conditions in the world. The most obvious application was harnessing the energy produced in reactors for generating electric power.

That application was immediately recognized as an opportunity by industrialists, notably by DuPont's Crawford Greenewalt, on December 2 after witnessing the pile go critical. It was not coincidental that DuPont, ensconced at Hanford, later led the development of large-scale plutonium-producing reactors. Nor was it just chance that the chemical firm Monsanto managed Oak Ridge during the war years. In ensuing years, Monsanto applied chemical techniques it had participated in at Oak Ridge to its burgeoning product lines.

Other applications were to medical technology, especially nuclear medicine. The use of radioactive isotopes in cancer treatment had begun before the war but now gathered speed. And new techniques were arising, including the use of protons produced in particle accelerators to reduce solid tumors. The innovator in this work was Bob Wilson, formerly the head of the research division at Los Alamos, whose 1946 paper on nuclear medicine in the journal *Radiology* was seminal. Wilson had experienced both positive and negative emotions about his association with the Manhattan Project, but he now turned the experience into a beneficial one.

Wilson focused on basic research for several years and then, in 1976, having proved his administrative skills in Los Alamos, was appointed director of the National Accelerator Laboratory, later renamed Fermilab, less than an hour's drive from Chicago. In 1996, during one of the last talks he gave, Wilson reminisced:

> At Los Alamos, where I had been working for the past five years, no matter how justifiable it may have been, we had been working on one thing, and that was to kill people. When that became crystallized in my mind by the use of the atomic bomb at Hiroshima, it was a temptation, to salvage what was left of my conscience, I suppose, and think about saving people instead of killing them. I jumped into the most obvious thing I could do next: because one could hurt people with protons, one could probably help them too.

Not all Los Alamos scientists shared Wilson's desire to make amends. Some wanted to continue along the path of building an arsenal of weapons or, like Teller, explore developing even bigger bombs. Many gravitated back toward university life. In the words of Hans Bethe, "We all felt that, like the soldiers, we had done our duty." Of the many who were immigrants, like Fermi, they had obeyed the American oath of citizenship that reads, "I will perform work of national importance . . . when required by the law." Undoubtedly they, with their American compatriots and British collaborators, had performed work of "national importance." But their work also contributed to the nuclear dilemmas that persist to this very day.

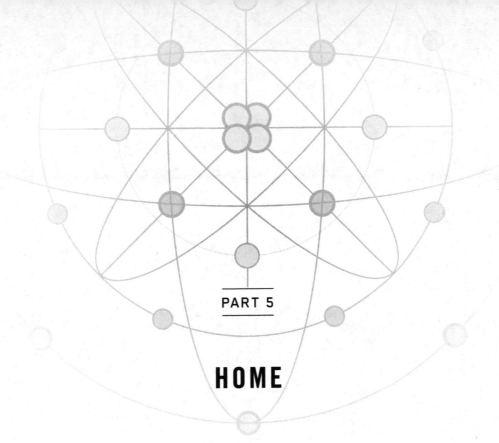

PART 5

HOME

35

PHYSICIST WITH A CAPITAL *F*

Hiroshima and Nagasaki acted as exclamation points for America's dominance as a world power. Already well on its way to becoming a major economic and political force, the young country of immigrants and refugees was now the undisputed leader in science. Enrico Fermi, at age forty-five, stood at its forefront, his new homeland's greatest physicist. He was, as succinctly described by one of his Italian peers, a physicist with a capital *F* (the Italian word for physicist being *fisico*).

Fermi's primacy did not hold in terms of being a physicist with political influence. Oppenheimer walked those corridors virtually alone, with ease and acumen. And Fermi's primacy was not true in terms of public image. Einstein, in his perch at the Institute of Advanced Study, had caught the world's imagination and adulation. But Chicago, home to multitudes of immigrants, claimed Fermi as its own. The city's newspapers celebrated him and other news media embraced his accomplishments.

Americans were still reeling from the magnitude and mystery of nuclear power. The Smyth report aimed to elucidate these complexities to them but it took America's most accessible medium to educate the public, not only about the history of the bomb but, perhaps more important, about how to harness nuclear energy for peaceful purposes. In a documentary entitled *The Quick and the Dead*, the NBC broadcast network tackled explaining the dangers and promise of atomic energy. The title was not welcoming, but the host of the series was. Listeners were riveted as Bob Hope, America's much-loved comedian, presented the four-week miniseries with a cast of familiar actors, including the famed Helen Hayes as Lise Meitner. Fermi's character was prominent. Neither politician nor icon, he had been popularized.

Fermi, never comfortable in the spotlight, tried to avoid it. Even when receiving the Nobel Prize in 1938, he did not like being the center of attention nor the event's pomp and circumstance. After World War II, Enrico was flooded by awards, honors, and medals that he graciously and modestly accepted. The impassioned debates that followed the dropping of the bomb drew him into the maelstrom of discussions about how nuclear energy should be applied and regulated in the future. Fermi's reputation inevitably made him part of controversies he would have preferred to dodge. But he could hardly do so, given that the bomb, as described by one politician, was "the most important thing in history since the birth of Jesus Christ."

In less flamboyant language, Major General Leslie Groves wrote to Fermi on September 28, 1945, thanking him for his war effort:

> Your scientific skill and judgment, your energy and ingenuity, and your self-sacrificing devotion to our cause are beyond praise. The forces of nuclear energy, which you helped so significantly to develop for use against the enemy, will, we pray, be wisely controlled in the days to come to ensure peace and to further the welfare of mankind. But no future events can dim the splendor of the results attained in the immediate past through our ability to make military use of these forces. For your indispensable part in this attainment, I thank you, on behalf of the War Department, as the agent of the American people.

While firmly embedded in the minds of a wide public, Fermi was also revered by "his own"—rank-and-file physicists, budding stars, and eager recruits alike. No one matched Fermi. People looked to him for instruction, guidance, and inspiration. A much wider audience was newly appreciating the role that a small group gathered on a Chicago squash court had witnessed four years earlier.

Part of Fermi's renown was due to the fact that the world had changed and physics was changing as well. The twentieth century's early decades, the era of revolutionary theories such as relativity and quantum mechanics, had been stunning but had passed. Now was the time for a continual interplay between experiment and theory, for interpreting clues offered by powerful scrutinizers of both the microscopic and the macroscopic. The thrust of these scientific endeavors was to make use of new tools and to think big. Government funding was flowing at unprecedented levels, forcing decisions on how best to utilize those monies.

Fermi accurately sensed the changing landscape and found it much in accord with his natural inclinations. His most famous contributions to physics were not leaps into the unknown. In the case of neutron bombardment, he had simply been searching for a projectile that could penetrate the nucleus unperturbed. As for Fermi's weak interaction theory, it came by looking for how quantum field theory could explain the nagging experimental question of missing energy in beta decay.

In moving forward, Fermi believed he was continuing a natural progression rather than attempting to embark on a revolutionary path. Ernest Rutherford had discovered the atomic nucleus in 1911 by observing the deflection of alpha particles bombarding a thin gold foil. Using a more powerful beam, James Chadwick had uncovered the neutron in 1932. It had been clear to Fermi since the late 1930s that the development of the powerful particle accelerators known as cyclotrons would cause that frontier to move forward. It had become possible to probe matter on smaller scales, to go deep inside the nucleus and hopefully uncover the nature of the mysterious forces holding neutrons and protons together inside it.

Fermi's incremental approach to physics kept him far more relevant than any of the great theoretical physicists he had admired in his youth. Now physics was looking to experiments for indications of its future and Fermi was the only physicist to combine being a great experimentalist with being a great theorist.

The older Einstein, Bohr, and Schrödinger, still esteemed, were well past their productive prime. To a large extent, the dominant theorists of Fermi's generation, such as Dirac, Heisenberg, and Pauli, had remained in familiar territory, still searching for an equation that would explain the behavior of nuclear constituents. By the postwar years Dirac had become, as his recent biographer characterizes him, "an ambassador of mathematical beauty." Heisenberg, who had been touting his unified field theory for years, finally unveiled it with great fanfare in 1958. It was viewed unfavorably by almost everybody, including Pauli, who had initially collaborated with him. His approach seemed mired in the past.

The next wave of leaders in physics was not coming forth from Göttingen, Munich, Leipzig, or Zurich, but from American academic institutions such as Berkeley, MIT, Princeton, and Columbia. And more than anywhere else they appeared at the University of Chicago, in America's heartland, following a trail set by Fermi. As one scientist commented, "Fermi was the Pied Piper who brought them."

The Chicago physicists were eager to attack a formidable problem that had surfaced but not been resolved in the 1930s: what holds nuclei together and why do they sometimes split apart? In the miracle year of 1905 when he proposed his theory of special relativity, Einstein had also envisioned the electromagnetic force as carried by particles. They would come to be known as photons, particles with no mass. In 1935, the Japanese physicist Hideki Yukawa said that the nuclear force might be carried by an altogether different kind of particle: a meson.

Quantum field theory provided a description of mesons being exchanged back and forth between the neutrons and protons in the nucleus. Mesons appeared and disappeared in a flash. Observing a meson was a problem, because Yukawa's theory asserted they had a large mass. The mass of a meson was estimated to be approximately 15 percent of the mass of the proton. The cyclotrons of the 1930s offered no hope of directly detecting mesons, since these accelerators could not provide enough energy to produce such a massive particle. However, advances in accelerator technology in the postwar period made possible what had previously been impossible. A much enhanced version of the cyclotron, referred to as a synchrocyclotron, boosted particle energy significantly.

Building such a machine would be a major engineering effort. Fermi was extraordinarily eager to proceed, anticipating that it would lead to a wave of discoveries once it was operating. The confirmation of Yukawa's conjecture in 1947 added impetus to its construction. A team from Bristol University had observed his meson in a cosmic ray. It was given the name pi-meson, or, more simply, pion.

Almost simultaneously, a high-level government official came to Chicago. Such was the faith in Fermi's record of achievements that the official only inquired how *much* funding Fermi needed, not to what purpose. Hearing about the prospective grant, an excited Herb Anderson asked Fermi, "What would you like? I'll build it. An accelerator, a big computer. . . . You name it." Fermi chose the accelerator, and Anderson, joined by John Marshall, began building it.

Originally scheduled for completion in May 1950, the complicated machine did not start operating until February 1951. By April of that year it was running at full power. The usually calm Fermi was beside himself, Laura noting that her husband was "as pleased as a child who has received a new toy long dreamed of and exceeding expectations. He played with the cyclotron at all hours of day and evening during that summer of 1951. He allowed the cyclotron to upset his routine."

Fermi had certainly not been idle while waiting for the completion of his "new toy." He had carried out an important set of neutron experiments with Leona Woods Marshall and others at the Argonne reactor and, in a visionary paper, had shown how intergalactic magnetic fields could provide the mechanism to accelerate cosmic rays.

Fermi had, however, always been thinking about what experiments he might perform once the machine was ready. It would provide not just a few pions, but a whole beam of them. The beam's energy could be varied and the pions directed toward a target of protons, subsequently making it possible to observe the angles at which the pions were scattered. The pion-proton interaction and therefore the nuclear force would become known to a previously unobtainable degree.

When the accelerator started operating, some unexpected results in the scattering experiments appeared. Fermi had not anticipated them but was prepared for surprises. There were hints of new particles being produced. These indicated that the search for the ultimate structure of matter would be lengthy. Perhaps one day it might even be shown that pions were composites of other particles, as Fermi had conjectured in a 1949 article written with his former student Chen Ning Yang. More powerful particle accelerators would be built in the coming decades, but Fermi was convinced that the experiments he was carrying out were the gateway to making progress.

The scattering experiments were also ensuring progress in a different direction by calling for development of novel computational tools. Fermi's office desk calculator and his constant companion, a small slide rule, were no longer sufficient for analyzing the enormous amounts of data being produced. But computers, first introduced by the late 1940s, could perform the task. They were tools of an entirely new character and Los Alamos had one. It would prove valuable for studying the details of the thermonuclear reactions underlying the H-bomb's functioning. Now, in a completely different vein, it allowed Fermi to analyze his experimental results. Back in 1947, Anderson had asked him if he wanted an accelerator or a big computer. In 1952, he had both.

Looking beyond his immediate interests, Fermi also saw that computers provided a way to deal with previously intractable physics problems, ones of a nonlinear kind in which the output is not simply proportional to the input or in which the expected approach to equilibrium is not reached. Research along the lines that Fermi conducted at Los Alamos with his mathematician friend Stan Ulam and others during the summers of 1952 and 1953 is seen as pioneering work toward contemporary fields, such as chaos theory.

Although advancing science was always Fermi's driving force, he was optimistic that the progress along these paths would eventually benefit society as a whole. In fielding questions such as "What does it matter?" about the development of the new tools, he answered as follows in a 1952 speech at the University of Rochester:

The history of science and technology has consistently taught us that scientific advances in basic understanding have sooner or later led to technical and industrial applications that have revolutionized our way of life. It seems to me improbable that this effort to get at the structure of matter should be an exception to this rule.

Remembering also how physics had contributed to the development of weapons, Fermi concluded with a warning for his audience, "What is less certain, and what we all fervently hope, is that man will soon grow sufficiently adult to make good use of the powers that he acquires over nature." It is probably as close as Fermi ever got to imparting a moral and cautionary message regarding scientific advances.

36

THE FERMI METHOD

Fermi was delighted to get back to the teaching environment of a university campus after leaving Los Alamos. The quality of physics students at the University of Chicago was extraordinarily high. Suddenly physics, once a relatively dormant and obscure field, was attracting top students at both the graduate and undergraduate level. Dramatic discoveries, astonishing tools, and high-profile visionaries all added to its allure.

Fermi's reputation as a professor was legendary and well earned. One doctoral student, still tentative about his chosen field, recalled how "my first physics exposure at the University of Chicago was to Fermi's famous 8:00 a.m. introductory lectures on nuclear physics. By about 8:15 I knew I had made the right decision!" In homage to his mentor, Marshall Rosenbluth later became known as "the Pope of Plasma Physics." In 1985 he won the coveted Enrico Fermi Prize.

Fermi's lecture notes for his nuclear physics course, carefully crafted and painstakingly detailed, were so esteemed that three of his students decided to type them up, enter the formulas by hand, and distribute mimeographed copies. As word spread, in 1949 the University of Chicago Press photo-offset the lecture notes into a book on nuclear physics. By the time of the sixth edition, twenty thousand copies had been sold; their format was unchanged. Fermi's lectures on quantum mechanics and thermodynamics were also published, praised, and appreciated by physics devotees not privileged to have him as a classroom teacher.

Fermi had the further distinction of an astonishing record: six of his Chicago students and one from Rome won Nobel Prizes in Physics. To this day, the record is unmatched. All the more remarkable was the fact that these laureates represented the respective disciplines of theory and experiment. It was yet another testament to Fermi's command of both fields. The Pope was training a cadre of future leaders; many viewed Fermi's teaching as his greatest contribution to the physics of the postwar period. The students' feelings were reflected in a 1954 letter one of the Nobelists wrote to Fermi, "If I am to be regarded as a decent physicist, it is mostly because of your training."

Re-creating the atmosphere of Via Panisperna, Fermi regularly invited advanced students to small gatherings in his office with its adjacent experimental laboratory. The setting reflected his partiality for simplicity and functionality. There were no comfortable sofas or rows of book-lined shelves in the office, only a blackboard, a plain metal desk, some chairs, and a few filing cabinets that stored stacks of notebooks. These notebooks, carefully annotated, carried a wealth of knowledge. A self-confessed nonreader of traditional texts, Fermi referred to these notebooks as his artificial memory, relying on them for information or a specific formula.

Office discussions ranged from abstract Riemannian geometry to the practicalities of electric noise in circuits. Fermi insisted that every assumption be tested and no formula taken for granted. One of the most brilliant theory students claimed they all learned from Fermi the maxim "Physics is to be built from the ground up, brick by brick by brick, layer by layer."

Never seeming to lose an opportunity to teach and learn, Fermi also mixed with students by eating lunch with them and a few colleagues at a long table in the main campus dining hall. A spirited discussion was guaranteed. The atmosphere was casual, far from European formalities where "Herr Professors" reigned.

Aside from the world of work, Fermi extended his relationship with students to the world of play. He joined them in their outdoor activities and sports, displaying his considerable endurance and his well-known desire to win. The short and less-than-graceful Italian was considered a ferocious tennis player who would run and hit in the midday sun until he thoroughly exhausted any opponent. Harold Agnew— who had worked with Fermi on the pile in 1942, followed him to Los Alamos, and returned to Chicago as a Ph.D. candidate—could attest to his mentor's exuberance on the tennis court, in the waters of Lake Michigan, and, most improbably, on the dance floor.

Square dancing was an import of the Fermis from Los Alamos days and was a frequent form of entertaining colleagues, graduate students, and occasionally undergraduates in the spacious third floor of the Chicago house. The dances were predictably lively, although bizarre to foreign students. "They were my first introduction to occidental culture," recalled the 1957 Nobel laureate Tsung Dao Lee, who had arrived in Chicago from China in 1946 as a twenty-year-old. "Enrico's dancing, Laura's punch and Agnew's energetic calling of 'do-si-do' all made indelible marks on my memory." The very essence of square dancing seems antithetical to the innovative and independent personalities of physicists. Dancers obey instructions, follow patterned footsteps, and stay within the confines of a square box. More typically, physicists tend to think independently and outside the box, beyond boundaries and constraints.

Maybe square dancing, this most American of gambols, provided some welcome respite and a semblance of normal life. Although Fermi considered himself "perfectly normal," one can argue differently. He saw the world through a special lens, that of quantification and rationality. In the physics community, the approach was so identified with him that it is referred to as "the Fermi method." Questions that he posed, ones that could be answered by careful order-of-magnitude reckoning, became known as "Fermi questions."

The idea was to arrive at an approximate answer, doing so, as one colleague said, "with a minimum of complication and sophistication." The Fermi method combined a breadth of knowledge, mathematical acumen, a strong dose of intuition, and mental agility. The Pope's accurate estimate at Trinity of the explosive power of the first atomic bomb was considered a classic example of the approach.

Numbers ruled and Fermi was a master at discovering them, interpreting them, and applying them. The joys of music and art seemed lost on Fermi, who was thought to be without much aesthetic appreciation. Quantification, not the ambiguities of aesthetics and taste, was what made him happy and secure. When his Chicago protégée Leona Woods Marshall took Fermi to the Art Institute to view a portrait exhibit of immigrants, she hoped he would identify with these fellow new Americans and appreciate the artistic treatment of them. It did not take long for him to pull out his omnipresent slide rule, compute the ratio of their body to leg lengths, and happily conclude that his own measurements coincided with the center of the distribution he had just calculated.

Other times—while hiking, for example—Fermi did not rely on a slide rule and used his thumb as a yardstick, placing it in front of his left eye while closing the right, thereby accurately calibrating mountain range distances, heights of trees, and speeds at which a bird flew. To Laura's somewhat amused consternation, one of his favorite pastimes was classifying people, on a scale from one to four, according to intelligence. When during their courtship Enrico gallantly offered to share the classification of four with her, Laura rejoined that if he was allowing her to be a four, he must have intended for there to be a scale of five. Five was reserved for Fermi alone.

What saved Fermi and his obsession with numbers from being overbearing was his amazing accuracy and his sense of humor. When Fermi thought, he played. This playfulness spilled over into his teaching, whether of undergraduates, graduates, or his peers. And it surfaced in sundry ways.

To encourage students to develop their skill in estimating orders of magnitude, a sort of mental calisthenics, Fermi would pose seemingly outrageous questions. Once, pointing to the university's lunch hall windows, he asked his students to estimate "How thick can the dirt be on the window up there before it falls off?" Another time he asked, "What is the number of sheep in Nevada?" His most cited question was

"How many piano tuners are there in the city of Chicago?" In the process of answering these questions, easy or hard, the students gained confidence, coming to feel, as one put it, that "we could solve any problem."

Fermi seemed to have an uncanny ability to rapidly grasp a problem's essential features. His graduate students would joke that his legendary intuition was only possible because "Fermi had an inside track to God."

Fermi's method was not limited to students. Colleagues treasured his give-and-take, consulting him even when the subject was far from his expertise. However, when it came to serious personal problems, Fermi was "certainly not the person to go to." A colleague describes him as reticent about talking about feelings, "not cold, but not warm either." Fermi was at his best as a scientist.

A 1983 Nobel Prize winner, the very distinguished Chicago astrophysicist Subrahmanyan Chandrasekhar, collaborated with Fermi on important problems in astrophysics regarding galactic magnetic fields. He described the impression Fermi made on him as being akin to that of "a musician who, when presented with a new piece of music, at once plays it with a perception and a discernment which one would normally associate only with long practice and study."

Maria Goeppert Mayer, another Chicago colleague, acknowledged in her 1963 Nobel Prize acceptance speech that her research had been turned around by a casual question Fermi posed to her about a possible interaction affecting the structure of large nuclei. Mayer was the second female laureate ever in physics, preceded only by Marie Curie. What Mayer did not know was that ten years earlier, when a prominent Harvard professor of chemistry wrote to Fermi inquiring about her, the professor started his letter, "I had the misfortune of having been made a member of a committee which is recommending a woman scholar for a distinguished appointment." Fermi responded the next day, ignoring the gratuitous gender reference. "I believe that Maria G. Mayer is without any [reservation] an outstanding scientist," he wrote, adding that her work was "doubtlessly the most important contribution to the physics of the nucleus in the last five years."

Given the reputation Fermi had acquired, physicists came from afar to benefit from his opinion on their work and seek his advice. The great theorist Freeman Dyson described an example from the spring of 1953. Then a young but already well known Cornell faculty member, Dyson traveled to Chicago to show Fermi a series of calculations he had undertaken with his graduate students. Their results matched the data from the pion-proton scattering experiments Fermi had performed with his own students. But Fermi was skeptical of the significance of such calculations. In a 1951 presentation for the twentieth anniversary of the founding of the American Institute of Physics, Fermi had stated that in attempting to understand the subject, "we must be prepared for a long pull." He did not estimate how long that pull might be. It would be another quarter of a century before quarks and gluons provided an answer.

When Dyson met with him in 1953, Fermi welcomed him politely, but he quickly put aside the graphs he was being shown indicating agreement between theory and experiment. His verdict, as Dyson remembered, was "There are two ways of doing calculations in theoretical physics. One way, and this is the way I prefer, is to have a clear physical picture of the process you are calculating. The other way is to have a precise and self-consistent mathematical formalism. You have neither."

When a stunned Dyson tried to counter by emphasizing the agreement between experiment and the calculations, Fermi asked him how many free parameters he had used to obtain the fit. Smiling after being told "Four," Fermi remarked, "I remember my old friend Johnny von Neumann used to say, with four parameters I can fit an elephant, and with five I can make him wiggle his trunk." There was little to add. In retrospect, Dyson was grateful for Fermi's rather brusque judgment, saying it "saved me and my students from getting stuck in a blind alley."

The Fermi method was not always applicable and sometimes could backfire. Talking once with a group that included Leo Szilard, Fermi wondered how one might determine whether there were extraterrestrial beings. What were the possibilities? After estimating the number of stars in the universe, the number of planets around them with liquid water and the evolution of some semblance of human life, Fermi asked if there could be a civilization that could communicate with Earth or even colonize it. If so, he rhetorically queried, "Where is everybody?" Szilard quickly responded, "Enrico, they are already among us. We just call them Hungarians." He was referring to the almost simultaneous appearance on earth of Teller, von Neumann, Wigner, and of course himself.

One of those Hungarians, the bushy-browed Teller, whose associations with other scientists were often volatile or prickly, was on good footing with his Italian co-immigrant. They taught a seminar together in Chicago, Teller often faltering in his explanations. Fermi would step in, remarking, "What Edward is trying to say is," and then articulate the concepts in his own lucid way.

By the end of the 1940s, their Chicago interactions were becoming far more rare. Teller was spending more time working in Los Alamos. The Super was once again becoming paramount and he was fervently promoting it. "What Edward is trying to say" was crystal clear. The zealous scientist, believed to be the inspiration for the movie *Dr. Strangelove*, was convinced America would be safe only if it succeeded in making a hydrogen bomb.

37

THE SUPER

The debates about further exploration and use of nuclear energy were hotly pursued not only in the streets and homes of Americans but also in the corridors of Washington. The atomic bomb was not an apparition; it was here to stay. Security issues and military applications overshadowed moral and ethical considerations. How many bombs could or should be built? Could they be made bigger? How soon would other countries possess the technology to make bombs?

During the immediate post–World War II years, United States government circles pondered how nuclear weapon development and nuclear power management should be regulated. These concerns were intensely argued as the Atomic Energy Act made its way through Congress in 1946. Several physicists, Szilard and Oppenheimer in particular, had tried to influence its formulation. A major bone of contention revolved around technology transfer—that is, sharing research with other scientists and countries. The act in its final form specified that all information regarding nuclear weapons was classified unless specifically indicated otherwise. Given the increasing hostility between the United States and the Soviet Union, attempts to either ban nuclear weapons or have an international control system were tabled.

Another contentious question was the proposed shift of all atomic energy production facilities from military to civilian control. When President Truman signed the act into law in August, the Manhattan Project basically morphed into the Atomic Energy Commission (AEC). Los Alamos was no longer under the aegis of the army. Instead, it was operated by the University of California in a contractual agreement with the AEC. A new director, Norris Bradbury, a Stanford University professor who had been at the laboratory since 1944, became Oppenheimer's successor.

A provision of the AEC, a civilian board of five members, was that it was to consult regularly with a General Advisory Committee of nine scientists. When it was effectuated on January 1, 1947, Fermi was named to serve on the GAC. The president had asked him and he, as a good citizen, felt he could not refuse. Fermi anticipated that his appointment would be relatively apolitical, focusing on technical questions. Instead, it would force him to take a public stand on one of the most perilous questions the world could imagine.

Fermi assiduously tried to avoid the political arena. Writing in January 1946 to Edoardo Amaldi, Fermi mocked those physicists who "are much more occupied with politics rather than science and are spending their time in pleasant conversations with senators and congressmen." Divisions among scientists characteristically ran deep. Fermi skirted around the personality clashes between Teller and Oppenheimer as well as the national controversies between advocates and opponents of weapon proliferation. The most astonishing fact is that physicists on both sides of a given issue seemed to universally like Fermi, forgiving and excusing his neutrality. In the words of Norris Bradbury, "No man could have been more respected for his achievements, nor more loved for himself."

The GAC's first meeting took place in Washington on January 3. The group—almost all of whom had been involved in the Manhattan Project—chose Oppenheimer to chair the committee. They also decided not to keep minutes of their discussions; at the end of each meeting they would submit a set of recommendations to the AEC.

As their first task, the AEC decided to assess the nuclear stockpile in Los Alamos. On a site visit, commission members were shocked by what they found there. No nuclear weapons were ready for use, an especially frightening realization since they had assumed, as had the president, that America's nuclear stockpile was the deterrent to the Soviet Union's far greater land army in Europe. The laboratory reorganized and developed a formidable nuclear arsenal over the next few months.

A topic that hovered over the AEC was whether to initiate a crash program for building a hydrogen bomb, or H-bomb, alternatively called the Super. Based on fusion rather than on fission, the Super was potentially thousands of times more powerful than the A-bomb. The prospect had obsessed Edward Teller since Fermi had casually introduced it during an after-lunch stroll in September 1941. Fermi had been musing whether fission could re-create on earth the conditions present in the sun's core, reaching temperatures above 15 million degrees Centigrade, sufficient for fusing hydrogen nuclei into helium ones. The idea was fleeting for Fermi, but Teller fixated on it.

The concept had been discussed the day after the Nagasaki bombing by the same scientific panel that had given advice on whether or not to drop the A-bomb. The four panel members feared that once the military heard of the possibility for a more powerful bomb, they would want to pursue it. The panel were not in favor of doing so but felt they needed to pose the question. In a set of recommendations, they wrote that prospects for developing the Super were "quite favorable." But panel members continued by asserting, "We believe that the safety of this nation—as opposed to its ability to inflict damage on an enemy power—cannot be wholly or even primarily in its scientific or technical prowess. It can be based only on making future wars impossible." They were unanimous in their belief that possessing the weapon would not protect the country from future war.

The matter was largely left open afterward; it was still unknown whether such a formidable weapon could actually be built. With the formation of the AEC, the idea of the Super surfaced again. The commission's first chairman, David Lilienthal, a progressive attorney known for leading the Tennessee Valley Authority, regarded such a prospect as obscene, a form of technological fanaticism. Given that such a bomb would be far more destructive than one employing a fission mechanism, he felt it could only be a weapon of mass destruction and therefore could not be justified on moral grounds. Lilienthal wanted nothing to do with it.

Pressures mounted on September 23, 1949, with Truman's announcement of the atmospheric detection of a Russian nuclear bomb explosion. Labeled Joe-1 as a nod to the Soviet Union's leader, Joseph Stalin, it suggested that America could no longer reign supreme over advanced weaponry and that stockpiling fission weapons could not be counted on as a sufficient deterrent to war. However, to the chagrin of the congressional committee charged with overseeing the AEC, Lilienthal did not put the Super on the AEC's October 5 agenda.

Incensed by what he perceived as Lilienthal's obstructionism, Lewis Strauss, one of the five AEC members, took matters into his own hands. Circumventing the regular channels of communication, he asked the secretary of the National Security Council to immediately take the matter up directly with the president. Truman was again surprised. He had not been told about the A-bomb's existence until after Roosevelt's death. Now, four years later, he once again had been left in the dark. On October 6, 1949, he was informed for the first time about another potential weapon—the Super.

These events triggered a flurry of meetings and memos. Should Los Alamos gear up for another crisis? Who would or would not join the effort? Teller, based in Chicago but with summers dedicated to H-bomb research in Los Alamos, had already decided to spend the 1949–50 academic year at the laboratory. He now led the charge to develop an "all hands on deck" approach, meeting with key physicists such as Hans Bethe, James Conant, Ernest Lawrence, John Wheeler, and Luis Alvarez, all previously involved with the A-bomb. He argued for a crash program for the H-bomb, a rebirth of the intensity afforded the A-bomb. A new Manhattan Project was essential for survival, according to Teller.

Due to coincidental luck, at least in his mind, Fermi had escaped the turmoil. He had left for Italy in late September on a long-planned visit, his first since the war. However, he was expected back in the States in time to attend the GAC's critical next meeting, scheduled for October 29 and 30. After his eighteen-hour flight, none other than Teller greeted an exhausted Fermi at the Chicago airport, urging him to come immediately to Los Alamos and join the H-bomb push. Fermi, regarding his colleague as close to maniacal on the subject, was non-committal. For the moment at least, his sights were focused on getting to the GAC meeting in Washington.

It became clear at the late-October meeting that none of the GAC's members was in favor of a crash program for the Super, although their degree of opposition varied. Conant was flatly opposed on moral grounds, while Fermi and Rabi were intrigued with the scientific feasibility of developing such weaponry. But the argument that it was necessary to have an H-bomb for defense and retaliation purposes against the Soviets held no sway with any of them. An arsenal of conventional nuclear bomb bombs, the GAC argued, would be more than enough.

While essentially agreeing among themselves, GAC members could not come to consensus on the language of their recommendations, so the report has two annexes. Each annex includes references to genocide, a powerful word introduced in 1944 to describe crimes against humanity. Fermi and Rabi unexpectedly penned the stronger of the two annexes, voicing firm opposition: "The fact that no limit exists to the destructiveness of this weapon makes its very existence and the knowledge of its construction a danger to humanity as a whole. It is necessarily an evil thing in any light." Their annex concluded by appealing to the president "to tell the American public, and the world, that we think it is wrong on fundamental ethical principles to initiate a program of the development of such a weapon."

Shortly after Fermi returned to Chicago from the meetings, Leona Woods Marshall stormed into his office. She had heard of Fermi's position on the Super and could not fathom how he, as someone who always sought knowledge, "could have voted against finding out whether the hydrogen bomb worked." Confronted by one of his closest colleagues, a woman haranguing him about not proceeding with devastating weaponry, Fermi "flared up," according to Woods Marshall, leaving both of them "shaking and speechless." She had witnessed Fermi's occasional outbursts, recalling his childhood nickname of Little Match, but had not yet borne the brunt of his temper. For months afterward, the two of them stepped gingerly around each other, the wounds of their battle over the Super slowly healing.

When the AEC took a vote on the issue, it was three to two against developing the Super. But Congress and the Joint Chiefs of Staff countered the recommendations of the GAC scientists and of the AEC vote. The controversy was stoked further on January 24, 1950, by the British Secret Service's arrest of the German refugee Klaus Fuchs on suspicion of being a Soviet spy. Fuchs had not left Los Alamos until August 1946 and had been present during all the initial Los Alamos discussions about building an H-bomb. Could he, a brilliant physicist, have shared this possibility with the Soviets? Might they have forged ahead in the direction of the Super?

A few days later, on January 31, 1950, President Truman made an announcement applauded by most politicians but criticized by the majority of scientists. He directed the AEC to work on the Super. Teller felt triumphant. In addition, Truman placed a gag order on the GAC scientists, forbidding them to speak publicly about the details of the decision. A veil of secrecy once again descended.

Fermi found this exasperating. He had opposed secrecy on neutron research in 1940, bowing to pressure as he saw war becoming imminent. Then, there was at least a good reason for silence. But the postwar constraints of secrecy, far less evident, weighed heavily on him. This had struck him particularly in the fall of 1946 during the course of a visit by Edoardo Amaldi, his dear friend and collaborator in Rome.

Fermi had forewarned Amaldi about the issue of secrecy months earlier, writing him that government funding for science carried "some very serious inconveniences. The most serious of all are military secrets. In this regard one hopes that a good part of the scientific results that are still kept secret can soon be published, but for the moment things are proceeding very slowly."

Despite the warning, their meeting was painful. Amaldi later reflected on what it had been like. "We used to talk and talk, and it was quite clear that after the war he could not say everything any more. With another person, it could have been different, but with Fermi it was terrible. I don't blame Fermi—of course it was the situation."

Seeing the shackles of secrecy extended had annoyed Fermi, but the Cold War had come to Washington and a mood of fear prevailed. It did not help that Senator McMahon, the chair of Congress's Joint

Committee on Atomic Energy, was saying that war with Russia was inevitable and that the United States needed to "blow them off the face of the earth before they do that to us."

But could an H-bomb be built? Teller had proposed a scheme for the Super, but there were still numerous uncertainties. Hans Bethe, although opposing the Super, commented that nobody blamed Teller for thinking in 1946 that it might be feasible, but "he was blamed at Los Alamos for leading the Laboratory, and indeed the whole country, into an adventurous program on the basis of calculations which he himself must have known to have been very incomplete."

Teller's forte, as Bethe knew from his wartime experience with him, was not elaborate modeling. But in the summer of 1950, the Super's most ardent advocate had the help of three all-stars: the master calculator Fermi and the adroit mathematician Stan Ulam in Los Alamos, and the great John von Neumann in Princeton. At the end of the summer, all three concluded the bomb would not work. The prerequisite materials were scarce, the design faulty, and the will to proceed shaky.

Fermi thought the issue had been laid to rest, but the next summer, on June 16 and 17, Oppenheimer asked him to join a meeting at Princeton's Institute for Advanced Study, which he now directed. A summit conference of sorts, the meeting was intended to exchange information about a Super; it brought together key figures in the government, including AEC officials, and Los Alamos laboratory leaders. Attending as well were what the young physicist Ken Ford referred to as "the big three" consultants, Bethe, Fermi, and von Neumann. To everyone's shock, Teller and Ulam had produced a new approach making it likely that a Super could be built. The prospects were so promising that those participating at the Princeton summit, including Fermi, endorsed the project. For Woods Marshall, who had berated her colleague so strongly about his former position, this stance was more in keeping with the Fermi she knew.

In later explaining his reversal from the conclusions reached in October 1949, Oppenheimer said, "It is my judgment in these things that when you see something that is technically sweet, you go ahead and do it and you argue about what to do about it only after you have had your technical success." The morality of building such weapons had not changed. Oppenheimer justified his change of mind on other grounds. The Korean War had increased the tension between the United States and the Soviet Union. In addition, the possibility of only the Soviets' possessing thermonuclear weapons now seemed more ominous, more likely to upset the tenuous balance of terror. For Fermi as well, these were convincing reasons for changing his mind. Again and always the rational scientist, he felt that "once basic knowledge is acquired, any attempt at preventing its fruition would be as futile as hoping to stop the earth from revolving around the sun."

Both the United States and the Soviet Union did indeed pursue the development of the H-bomb. Los Alamos again swung into action, not at full throttle as for the Manhattan Project, but with a concentrated and intense effort. Fermi, who spent two months at Los Alamos during the summers of 1951, 1952, and 1953, continued to be an invaluable consultant on both developing the bomb and deciphering the results of Soviet explosions.

The Americans conducted their first full test of the Super on November 1, 1952, on a Pacific atoll. Its yield was more than 450 times as great as the bomb dropped on Nagasaki. Soon after, on November 24, the Executive Director of the Joint Committee on Atomic Energy wrote to Fermi asking for his personal estimate of how much lead time the United States had over the USSR. Fermi answered on November 26, "It is impossible to make anything but a wild guess of the answer . . . My utterly uninformed guess is that it might take them some time between two and five years." Again, Fermi's formidable intuition (*intuito formidabile*) was correct. Although ten months afterward the Soviet Union—basically for propaganda purposes—exploded a bomb with some elements of fusion, it was not until three years later that they tested a bomb truly like the Super.

—•——•——•—

38

CIRCLING BACK

Fermi had missed the October 5, 1949, General Advisory Committee (GAC) meeting and its arguments over weaponry. Instead, he was enjoying himself much more. He had arrived in Italy three weeks earlier, his first return to his homeland in the eleven years since he left in 1938.

Laura had gone back to Europe earlier, in the summer of 1946, to see her brother, her two sisters, their families, and various friends. By that time, she had completed the multiple details associated with selling one house and buying another. The Fermis, when they left New Jersey for Chicago, had thought the relocation would be temporary. Little did they suspect that yet another move—to Los Alamos—would occur. From 1942 to 1946, they had rented their Leonia house to a family, the Ferralls, who regularly informed them about house problems. Several letters from Mr. Ferrall, often beginning with "Mrs. Ferrall has noticed," recited a list of woes: the lack of screens, moth infestations, and dying trees. The American dream of home ownership, which Laura and Enrico had so avidly pursued as new immigrants, had its downsides. They sold the Leonia house in early June 1946, remembering one last thing. When they had bought the house in 1939, unsure of what the future might bring, the Fermis had placed some of their Nobel

Prize money in a lead pipe to protect it from dampness. It was buried in the basement. They did not forget to dig the money up.

The sale of the Leonia house coincided with the Fermis settling into their gracious new home in the Hyde Park neighborhood of the university. Laura had undertaken the exhausting task almost single-handed and in July 1946 was more than ready to reunite with her Italian relatives, while Enrico, Nella, and Judd stayed in the States. That summer Enrico was busy with work and, as had become customary, he was scheduled to spend several weeks in Los Alamos. Laura had arranged for temporary help in the Fermi household with the addition of Enrico's devoted student Harold Agnew, his wife, and their two-year-old toddler. Fifteen-year-old Nella, according to Agnew, did most of the cooking that summer. Unlike her mother, whose privileged childhood had left the culinary arts to servants, Nella was showing her competence and self-sufficiency.

Three years later, Enrico went back to Italy as well. It was an emotional trip for him. Although he was firmly embedded in America, Italy would forever be close to his heart, the country where his roots were. He had seen his homeland go astray and then be devastated by a war in which he fervently hoped Mussolini's government would be defeated. But he had also been apprehensive about what might happen there in the immediate postwar period. Accordingly, in a rare act of self-motivated political concern, Fermi had written in October 1946 to the United States secretary of state James Byrnes regarding his worry that in Italy "forces that tend to make the situation unstable are dangerously strong."

Though this was unstated, Fermi had clearly been troubled that the Communist Party, then part of the coalition government, might take control of the country. Fermi modestly added that he doubted this threat would be news to Byrnes, "but my knowledge of Italy where I was born and where I lived until seven years ago may make it of some interest to you to hear my views" and urged the secretary of state to "encourage the democratic forces in Italy." In the same letter, he made a plea for ascertaining the fate of Italian Jews who had been rounded up in 1943, deported to Germany, and never heard from. That referred to—among others—Laura's father. Fermi concluded his letter by saying, "I believe that any help that the United States can give Italy . . .

would strengthen enormously the chances that Italy may settle in a stable democratic form of government."

When Fermi went back to his homeland in 1949, he felt the gravest danger was over. Though Italy would not be admitted to the United Nations until 1955, it had returned to relative stability, making Fermi eager to support its rebuilding of physics.

His first Italian stop was the tranquil retreat of Como, showing no scars of the brutal war, its beauty preserved. Hillsides dropped steeply down, joining the thirty-mile-long lake, strikingly blue against a backdrop of the Alps. Memories flooded Fermi's consciousness when he thought back about the international Volta conference, a turning point in his career. Enrico had been thrilled to hear talks by iconic physicists of that time, Niels Bohr and Arnold Sommerfeld, the latter recognizing the young Fermi's promise. After faint acknowledgment on the Italian stage, Enrico's confidence had been seriously boosted by this explicit endorsement on the international stage. Sommerfeld's approval had been reinforced many times over by Enrico's extraordinary contributions to physics.

The 1949 conference focused on cosmic rays, of considerable interest to Fermi and an Italian specialty in the immediate postwar years. The literal high point of the conference was a special excursion to the new Italian cosmic ray observatory located on a high plateau between the Cervino (the Italian name of the Matterhorn) and Monte Rosa. Furthermore, reaching it required going through Champoluc, the village he and Laura had chosen for their honeymoon twenty-one years earlier.

The conference also provided an almost complete reunion of the Boys of Via Panisperna since Edoardo Amaldi, Bruno Pontecorvo, and Emilio Segrè were attending. This gave them a chance to discuss a nagging question. They, along with Franco Rasetti, had discovered in 1934 the effect of using slow neutrons to study nuclear structure. The Boys, at the urging of their mentor Orso Corbino, had filed an Italian patent on the technique they developed, subsequently extending the patent to other countries including the United States. After the war, slow neutrons had proved to be central to nuclear power production and had significant military applications. The patent was valuable.

Fermi, who had been the chief inventor, was the paramount figure in patent negotiations over just compensation. To his considerable annoyance, U.S. government lawyers asserted that Fermi's presence on the Atomic Energy Commission's GAC, even though unpaid, created a conflict of interest for him and the other patent claimants. Negotiations were drawn out and arduous. Fermi would have been content to let the matter drop. He had not taken patents out on any of his later discoveries at Columbia or Chicago. But he felt obligated to assert the rights of his Italian co-inventors of the slow neutron technique. And he resented the tactics of zealous lawyers who maneuvered their way around mandated patent laws.

The claim was further complicated in 1950 when the story took on a new twist. Pontecorvo mysteriously disappeared, apparently having fled to the Soviet Union. Since he had always seemed apolitical, this came as a total surprise. A patent claim being argued in the United States legal system was made even more convoluted by one of the claimants being a defector, a presumed Communist. After much legalistic wrangling, the claim was finally settled in 1953 with the distribution of modest compensations. A spinoff of the whole unpleasant process was Fermi's renewed disillusionment with politics, especially with serving on government advisory boards. On August 1, 1950, he declined reappointment to the GAC, a post he had held since January 1, 1947.

Once the Como conference was over, Fermi traveled to Rome to reunite with his sister, Maria, now a widow with three children. He learned more about the attempts she had made to save Admiral Capon and also how, during the horrendous Nazi raids of October 1943, she had hidden six Jews in her small house: two teenage boys, a mother with a small child, a middle-aged lady, and an old man. Enrico admired her greatly. He also took the opportunity to explain his involvement in making the bomb. Her issue, as she had articulated in a 1945 letter to him, was that it was dropped on innocent populations. Enrico seemed inured to the criticism and tried, with limited success, to minimize his role in that decision.

The highlight of Fermi's Italian visit was a two-week cycle of nine lectures, the first six delivered in Rome and the last three in Milan. His impending visit was publicized extensively and enthusiastically in the press and newsreels. The prodigal son was returning. Emotions ran high in the large audience attending his first talk on October 3 at the University of Rome's huge lecture hall, the Aula Magna. When Guido Castelnuovo, old and frail, rose to his feet to introduce Fermi, the animated crowd quieted.

Castelnuovo was the last survivor of the four great Rome mathematics professors who had offered a strong welcome to Fermi when he arrived in Rome in the early 1920s. The warm reception had been important for the young Pisa graduate's morale. And Castelnuovo had been a good friend of Admiral Capon, Fermi's father-in-law and a fellow Venetian: it was Castelnuovo's daughter Gina who had introduced her friend Laura to young Enrico.

During the Nazi occupation of Rome, Castelnuovo had barely escaped Capon's fate, doing so by hiding under an alias. In 1938, the four Roman mathematicians from that earlier day, and all other Jewish members, had been expelled from the century-old Accademia dei Lincei, the Italian society of notable intellectuals Mussolini had proceeded to abolish. With the war over, the Accademia was reestablished and Castelnuovo, pointedly, was chosen as its head. His task was to officially greet Fermi after an absence of more than a decade.

Castelnuovo's 1949 message to the physicists underscored that Fermi's visit was "demonstrating his interest in the young school of Italian physics to which he had given its first impulse." It was time to rebuild Italian physics. Rising to the occasion, Fermi gave a brilliant series of lectures on contemporary topics in physics. It was a remarkable display of omniscience and virtuosity. All the more impressive was that Fermi had worked in recent years on each of the topics, unfailingly opening windows to the future.

Fermi gave the last of his cycle of nine lectures on the twenty-first of October. His reentry to Italy had been a resounding success by any measure. But it also underlined his deep-seated Americanization, despite the country's flaws. He was happy to return to Chicago. Five summers would pass before he came back to Italy. Meanwhile, he resumed his yearly summer visits to Los Alamos.

The Atomic City on the mesa had changed in tenor. No longer an army post, temporary housing was being replaced by permanent structures. A state-of-the-art high school was built in 1946. Grocery stores were well stocked, and an ice cream parlor proudly advertised thirty-seven different flavors. It was beginning to look like a typical American town.

After the war, the laboratory had undergone a period of crisis. Most of the key personnel who had worked on the bomb had left, although many of them—like Fermi—had returned for summers. The lab had diversified somewhat, but its mandate to build an arsenal of nuclear weapons was primary. As the Soviet threat deepened and the Cold War took shape, this mission intensified.

Fermi was matter-of-fact about the laboratory's purposes. Unlike other elite scientists who had worked on the bomb, he did not subsequently join—much less establish—any antinuclear groups. Fermi was not opposed to weapon development, even though he advocated for further efforts at international control of nuclear weapons.

The year 1953, Fermi's last summer in Los Alamos, was bookmarked by two significant but seemingly unrelated occurrences. Toward its beginning, he had been once again reluctantly dragged into the political arena, elected to a one-year term as president of the American Physical Society. He expected the role to be largely ceremonial but, with America at the height of the McCarthy frenzy, he found himself repeatedly defending physicists against witch hunts. The best known of these began toward the end of 1953.

On December 21, J. Robert Oppenheimer, while he was director of the Princeton Institute for Advanced Studies—although still a government consultant—was shaken by the news that his security clearance had been suspended, presumably because of his earlier left-wing and Communist affiliations and his opposition to the development of the H-bomb. Confident of his loyalty to America and of his record of leadership, Oppenheimer requested hearings be held by the Atomic Energy Commission. He felt that otherwise his reputation would be tarnished. Fermi knew that were he called to testify at the hearings, he would do so unequivocally in Oppenheimer's favor. However, a colleague in Chicago who was having lunch with Fermi on the day they heard the news recalled Fermi's saying, "What a pity that they attacked him and not some nice guy like Bethe. Now we all have to be on Oppenheimer's side."

Although appreciative of Oppenheimer's contributions, Fermi—a self-made physicist from modest stock—was not a fan of the man he referred to as "born with a golden spoon in his mouth." The reasons for Fermi's coolness toward Oppenheimer are easy to imagine. The physicist Robert Wilson, who knew both, once compared the two. He described Oppenheimer as having a "personality crisis" every five years: "When I knew him at Berkeley he was the romantic, radical bohemian sort of person, a thorough scholar. Then at Los Alamos, he was the responsible, passionate leader that we all knew so well and was so effective. Later on he had another metamorphosis, becoming the high-level statesman who would call Acheson by his first name."

Wilson's description of Fermi conveys a very different persona: "Fermi, on the other hand, I never knew him to change, from the time I, as a student had seen him, to the time at Columbia, when I was involved in a collaboration with him, to the time at Los Alamos, to the time after that. It always seemed that Fermi was Fermi!" Oppenheimer and Fermi were both inspirational leaders, albeit in different ways. And, while certainly respectful, they were never comfortable with the other's approach to life or to physics.

The physics community reacted to the news of the hearings with shock. Oppenheimer was highly regarded by most of his peers, but within that group he had enemies, none more insistent and influential than Edward Teller. Teller had long been jealous of Oppenheimer's popularity and by 1954 wanted to "defrock him in his own church."

The hearings began on April 12 and lasted four weeks. Fermi's testimony, on April 20, expressed no ambiguity in his defense of Oppenheimer. Teller was another matter, giving damaging testimony on the twenty-eighth. At first he intoned, "I would like to see the vital interests of this country in hands which I understand better, and therefore trust more." When pressed, Teller asserted that "it would be wiser not to grant clearance." The ruling was announced by the AEC on June 30: Oppenheimer's security clearance was revoked. The physics community was aghast and largely ostracized Teller from then on. Their supportive feelings toward Oppenheimer were reflected in a quip by the rocket scientist Wernher von Braun: "In England they would have knighted him."

More was at stake than the fate of one man. Fallout from the hearings led to a new chapter on the relationship of scientists to government. Fermi had always sought to narrowly define his participation in government by focusing on technical matters. But others had quite comfortably bridged scientific and political spheres, seeing themselves as ordained to the "public-policy priesthood." That time had come to an end.

39

LAST GIFT TO ITALY

(*Ultimo Regalo all'Italia*)

In his role as president of the American Physical Society as well as on a personal level, Fermi was extremely upset by the Oppenheimer proceedings. America seemed to be taking a turn for the worse, McCarthyite hysteria conjuring up echoes of Fascist Italy. And he saw the emotional and physical price that the debacle had exacted from Oppenheimer. Adding to his distress was that Fermi regarded himself as a good friend of Teller's and bemoaned Edward's damning testimony.

Fermi welcomed the fact that coincidentally he had made plans to return to Europe in the summer of 1954. Luckily for him, Los Alamos would not be his usual destination; it was inevitable that conversations there would be dominated by what happened at the Oppenheimer hearings. He and Laura arrived in Paris on July 1 and spent a few days with their old Los Alamos neighbors Stan and Françoise Ulam in southern France. Ulam noted that his friend did not look well and was having trouble with his digestion. Laura thought it was probably due to overwork and tensions from the Oppenheimer hearings. She was hoping a break would do him good.

Fermi's plans for the summer were to teach at two physics schools, both times giving lectures to thirty or forty advanced students or recent

Ph.Ds. The first lecture was held at the Les Houches school, located in a village perched on a high Alpine meadow at the foot of Mont Blanc. The site afforded a magnificent view of the mountains and the whole valley of Chamonix, the busy resort center four miles away. The housing was in simple mountain chalets and the lectures were delivered in a converted barn.

The work schedule was intense but fortunately there was time for hiking. On the fourteenth of July, the French national holiday commemorating the storming of the Bastille, there were no lectures. Knowing of Fermi's interest in cosmic rays and his appreciation of the local scenery, the school's directors arranged for him and three others to have a special excursion to the French observatory near the top of the Aiguille du Midi, one of the needle-like peaks in the Mont Blanc massif.

The journey upward was an adventure: the present enclosed cable car lift had not yet been built. All that was available was a bucketlike carrier used for hauling supplies and construction workers. It was barely large enough for the four passengers crouched on its floor. As they entered the clouds, Fermi "smiled archly, saying he understood how angels felt." The Pope then "began singing the only hymn he knew: 'Mine eyes have seen the glory . . .'" Both the ascent and descent went smoothly, and the group returned safely to Les Houches.

Leaving the French Alps for Italy, the Fermis arrived on the eighteenth of July in Varenna on the familiar shores of Lake Como, where the second summer school took place. Today it is known as the Enrico Fermi International School of Physics. Meetings are held in the beautiful Villa Monastero, whose grounds are enchanting. A carefully tended formal garden has wide steps leading directly to the water's edge, inviting attendees for an afternoon swim. Fermi regarded it as perfect.

Fermi acted as teacher and critic. This was particularly important in the school's final lectures, devoted to the topic of future particle accelerators. Recovered from the ravages of World War II, European economies were beginning to thrive. The continent was preparing itself for a renaissance in science, with high-energy physics a paramount focus.

CERN, a symbol of European unity for nuclear research based in Geneva, was taking the lead. Edoardo Amaldi was one of its principal proponents and ardently spoke in Varenna about its progress. Hearing him, Fermi fondly remembered his good friend's visit to Chicago in 1946, when he offered Amaldi a professorship at the university. A serious lure had been the prospect of building a large cyclotron, the kind the two of them had envisaged in Rome years before. Now Europe was embarking on the quest to build bigger, more powerful accelerators. Fermi even joked about a giant one that would circle the entire earth, calling it a Globatron. Had he lived to see CERN's Large Hadron Collider, the world's largest particle accelerator, he would not have been altogether daunted by it or its discoveries of new particles, such as the Higgs boson.

But in 1954, accelerators were still relatively small, their future often the subject of rival national interests. Within Italy, the competition for a potential site to build an accelerator had been whittled down to either Frascati, just outside Rome, or to Milan. After Frascati was chosen, Pisa, an initial competitor, found itself with unallocated funds that had been earmarked. That is where Fermi came in.

When asked by leading Italian physicists how to spend the funds, Fermi advised developing a large computer. Having witnessed the computer's capabilities in Los Alamos, he considered this the wave of the future. Fermi also wrote a letter to the Rector of Pisa assuring him that such a computer "would constitute a means of research from which all the sciences and all avenues of research would gain in an inestimable way." This pivotal recommendation was referred to as Fermi's *"ultimo regalo all'Italia"* (last gift to Italy).

Afterward, the Fermis left Varenna and went to the Val di Fassa in the Dolomites. The Amaldis had preceded them for a vacation with their children. Enrico Persico, Fermi's childhood friend in Rome, joined them there. These environs held a special place in the hearts of all of them. This is where young Fermi had first met Laura, still a teenager. It is where on a winter vacation a decade later he had told the Boys about his new theory of the weak interactions, and where Ginestra Amaldi learned of her pregnancy. And it is where, in 1938, Fermi had sent letters to American universities saying he was ready to leave

Italy. Now, in middle age, they gathered again: the Fermis, the Amaldis, and Persico.

First came spirited doubles matches on the tennis court: the Amaldis against Fermi and Persico. The Amaldis were serious tennis players. The Fermis had even sent them a care package right after the war, containing much-appreciated tennis balls.

The history between Enrico and Edoardo had been meaningful, as was the friendship between their wives. Laura and Ginestra had met as young women, written a book together, and started families at almost the same time. By 1954, Ginestra was a well-known science interpreter and popularizer in the Italian media and had published a well-received book. She was delighted when Laura gave her a copy of the recently released *Atoms in the Family*.

The book, a memoir of Laura's life with Enrico, was admirable in its coverage of her husband's scientific contributions and their family life, while charming with its light touch and self-deprecating humor. Ginestra was particularly touched by its Italian memories, underlining the extent of the Amaldi-Fermi bonds. It described how Ginestra had organized a celebration for Enrico's Nobel Prize on the evening of the announcement and how the Amaldis, with Rasetti by their side, had been the last to wave goodbye to the Fermis at the Rome train station in 1938.

Their children had then all been very young. In the summer of 1954, almost sixteen years later, Amaldi's son Ugo was about to turn twenty and briefly appeared at the reunion of the families. Following in his father's footsteps, he was a physics student at the university. Connecting with the group in the Dolomites, he was desperately trying to follow conversations about physics and computers that Fermi, his father, and Persico were having. Similarly, his father, then a little younger than Ugo was now, had strained thirty years earlier to follow conversations about quantum theory between Fermi and his friends.

The Fermis' eighteen-year-old son, Judd, still called Giulio in Italy, was also in the Dolomite village, having joined his parents during the summer. Leaving the grown-ups, he went off on a six-day trek around the region's high peaks with his cousins, Maria's children, and Ugo Amaldi's fourteen-year-old brother, Francesco. They slept in mountain refuges and found the excursion a great adventure filled with animated conversations about literature and politics.

Handsome and a good head taller than his father, Judd would be going off that fall to start his junior year at Oberlin College in Ohio. Exceptionally bright, Judd was studying pure mathematics—a field that dealt with abstract concepts, far removed from math's more practical tools and techniques. It was not the kind of math favored by his father. Judd distanced himself as much as possible from Enrico's field and fame.

Judd's path through adolescence had been a rocky one, with tendencies toward depression and alienation. From Judd's viewpoint, his father had been both emotionally and physically absent during his childhood. Laura had hoped the Dolomite setting would bring her son and husband closer together, as well as give a boost to Judd's relationship with her. Unfortunately, this did not happen with any sustainability.

Fermi's European sojourn also included lengthy walks with Persico on the island of Elba, off the southern coast of Tuscany. The two Enricos hiked by themselves, recalling the rambles of their youth, when they shared common dreams and a mutual fascination with all things mechanical.

Persico described his dear friend on their last walk, "I found in him an old habit, that I think few knew, which perhaps will astonish those who knew him only superficially. Often, in moments of relaxation, walking or stopping to view a beautiful landscape, I would hear him reciting for himself long stretches of classical poetry, that since his youth he had kept in his memory as a rich treasure. With little inclination toward music, poetry took for him the place of song." Fermi may have seemed tone-deaf to his pianist friend Teller, but to his childhood friend Persico, he was melodious.

This most private of men seldom revealed himself to others. "Fermi was Fermi," in the words of his colleague Bob Wilson. But Fermi was also Enrico, a person with a depth of feeling that he rarely showed, even to his children. As his daughter, Nella, commented about her father, "It wasn't that he lacked emotions, but that he lacked the ability to express them."

40

FAREWELL TO THE NAVIGATOR

When Enrico and Laura returned to Chicago in September, they were exhilarated by their European journey, having stayed in places that were stunningly beautiful and meaningful to them. They had connected with treasured friends and enjoyed the receptiveness and appreciation showered on one of Italy's favorite sons. In spite of troubling health issues, Fermi had given an electrifying set of lectures and happily indulged in hikes, tennis games, and swimming.

Laura's reentry to the United States was greeted with accolades for *Atoms in the Family*. The University of Chicago Press had not expected it to be the bestseller that it became. When the *New Yorker* magazine published two excerpts from it under the title "That Was the Manhattan District—A Domestic View" on successive weeks in July, the book's popularity was clinched. It launched Laura on a successful writing career that covered such diverse subjects as Mussolini, Galileo, and recent immigrants who had contributed to America's greatness.

As soon as Fermi adjusted back to his Chicago routines, he made a doctor's appointment. Friends had been shocked at how thin and wan Fermi looked. The doctor told him there was nothing to worry about: it was psychological. For someone who was steady and unflappable, it seemed like a strange diagnosis. He chose to ignore any symptoms and, somewhat miraculously, continued a full schedule of lecturing. Concurrently, he fielded an enormous number of requests to accept honors, give guest talks, and participate in various conferences. He meticulously answered each inquiry, politely declining almost all of them.

It became unavoidable, however, that there was something seriously wrong with his health. Fermi was having more and more troubles with his digestion. In the beginning of October the diagnosis changed: either an esophageal obstruction or stomach cancer was the probable cause of his ailing. Exploratory surgery was scheduled for October 9 at the university's Billings Hospital. The operation showed that he had a widespread stomach cancer that had already metastasized. There was no hope: Fermi was told that he had only months to live.

Questions immediately arose as to whether the cancer might have been caused by exposure to radiation at Via Panisperna, Columbia, Chicago, and Los Alamos. It was an understandable reaction, believed by many, since Fermi continued to insist—throughout his career—on carrying out his share, or more, on experimental efforts. When Fermi worked on the synchrocyclotron, younger members of the group were concerned that he had received far greater exposure to radiation in his lifetime than they had. But Fermi was not to be deterred from participating in teamwork. There is no medical evidence, however, that his cumulative exposure led to stomach cancer. None of the other Boys died of cancer, nor did Anderson or Zinn, Fermi's two constant companions in building the pile. The incidence of Los Alamos scientists dying from cancer is similarly unremarkable.

As he faced the end, Fermi's personality remained unchanged, friendly but not effusive, rational in his judgments, often slightly ironic and in control of his emotions in an unforced way. Word spread quickly and many physicists close to Fermi made their way to his bedside to pay homage and say their farewells.

On the day after the surgery, Fermi was visited in his hospital room by his collaborator the Indian astrophysicist Subrahmanyan Chandrasekhar, known as Chandra, and Herb Anderson, who had been with Fermi through thick and thin, from Columbia days to Chicago days, with Los Alamos in between. They braced themselves, knowing that Fermi knew his death was imminent. Sensing they were at a loss for words and aware of Chandra's Hindu upbringing, Fermi quipped, "Tell me, Chandra, when I die will I come back as an elephant?" The conversation was easy after that.

When his young colleagues, Murray Gell-Mann and Chen-Ning Yang, visited him, Fermi calmly informed them of his condition and then pointed out by his bedside a notebook on nuclear physics. He hoped to edit it for publication in the time he had left. Yang remembers how "Gell-Mann and I were so overwhelmed by his simple determination and his devotion to physics that we were afraid for a few moments to look into his face." To another young colleague, Richard Garwin, who visited him, Fermi lamented the relative lack of public policy involvement in his life. This self-critique undoubtedly resonated with Garwin, who went on to have an extraordinarily distinguished career as a presidential adviser on science and security issues in addition to his notable contributions to basic physics.

Emilio Segrè flew to Chicago as soon as he could. Fermi had been his idol, his guide during all of his adult life. The Pope was infallible; he was smarter, worked harder, was stronger, had better eyesight, and could run faster. Nobody else's friendship and approval meant as much to him. How was it possible that he could be dying?

Emilio found Fermi lying in his hospital bed, serene about his fate. He was "being fed artificially. In typical fashion he was measuring the flux of the nutrients by counting drops with a stopwatch. It seemed as if he were performing one of his usual physics experiments on an extraneous object. He was fully aware of the situation and discussed it with Socratic serenity." To the very end, Fermi was busy calculating.

There were lighthearted moments in the conversation between the two immigrants. Fermi told Emilio how a Catholic priest, a Protestant pastor, and a Jewish rabbi had separately entered his room, asking for permission to bless him. The Pope added that he had willingly consented to each since "It pleased them and it did not harm me."

One of the touching moments recalled by Emilio was Fermi's pride in his wife's book. "I hope the book will be successful; it will help distract Laura from her grief for me. It comes at the right moment." Enrico lived long enough to see *Atoms in the Family* on the *New York Times* bestseller list. Ironically, once she was no longer in Fermi's shadow, Laura became a well-known personality in her own right.

As the two longtime friends, Segrè and Fermi, neared the end of their visit, Fermi—using a religious turn of phrase of his own—asked Emilio to summon Edward Teller, "adding with a slightly ironical smile, 'what nobler deed for a dying man than trying to save a soul?'" His view, as expressed to Emilio, was that "the best thing Teller can do now is to shut up and disappear from the public eye for a long time, in the hope that people may forget," but he still wanted to help his friend. On leaving the hospital room, Segrè was so upset he almost collapsed. He walked alone into a bar and ordered a stiff drink.

Teller was terribly nervous about visiting Fermi, who he knew had strongly disapproved of his negative testimony at the Oppenheimer hearings. The ensuing hostility of the physics community had been fueled by a July article in *Life* magazine and a subsequent book by two *Time-Life* reporters. They depicted Teller as a heroic figure responsible for the hydrogen bomb, having overcome opposition by Oppenheimer and the Los Alamos Laboratory.

Somewhat contrite about his overblown portrayal, Teller had drafted an article for *Science* magazine entitled "The Work of Many People." He had a copy with him and wanted Fermi to read it. After doing so, Fermi encouraged him to publish it, not only for its implicit truth—the bomb was indeed a product of many—but also to make amends. The article appeared in February 1955.

Stan Ulam, who came to visit the day after Teller, discussed with Fermi the uproar about their mutual colleague. Fermi told him of his attempt to "save a soul." Ulam was stunned by Fermi's ability to forgive and his almost superhuman serenity and calm. As the two discussed physics, Fermi offered an evaluation of his own work, telling Ulam he believed he had achieved two thirds of what he hoped to accomplish in his lifetime. His regret was that he could not complete the missing third. Like others, Ulam left the hospital room deeply shaken.

On November 3, Fermi came home to die. Laura had rented a hospital bed for him and he "told her to rent it only until the end of November because he wouldn't need it after that." Leona Woods Marshall, as a family friend as well as Fermi's collaborator, visited frequently during the next weeks, his last. When she attempted to talk about the two of them writing another paper, Fermi jested that it would have to include a black cross after his name, "directing the reader to a footnote that would say, 'Care of St. Peter.'" She drove home with "tears streaming down [her] face."

Everyone commented on how strong and dignified Laura was throughout the ordeal. Clearly, the marriage between her and Enrico had been a happy one, although being the wife of a physicist was not always easy. In her life with Fermi, she had assumed the position of a supportive, loving wife. This was reflected in her comments years later:

> Some physicists' wives believe that their husbands like physics better than they do their wives. And they may have a point. After work in the evening, when a wife is expecting a word of endearment like "I couldn't live without you," the husband is most likely utterly silent, absorbed in scribbling numbers and symbols on the margins of the evening paper. When she would like to go to the movies, he has a date with an experiment that cannot wait. There are other complaints, some more justified than others.

Laura's concluding line, "But all in all, life with a physicist is well worth living," understated the complications of that life and her love for Enrico.

In the years that they had been married, Laura gradually changed more than Enrico. As Bob Wilson had noted, "Fermi was always Fermi." In contrast, Laura was not always Laura. She began life as a conventional middle-class Roman and then, particularly after Enrico's death, shifted her sights.

She had played out her role as the wife of Fermi and mother of their children. Now she entered the arena of political activism. Less than a year after Enrico's death, she attended in Geneva the first International Conference on Peaceful Uses of Atomic Energy (August 1955), appointed to write an official account of American contributions to planning and procedures of the meeting. The conference steered her toward being a peace activist. Laura would also make her mark—well ahead of the times—on environmental issues and handgun control. Felled in 1977 by a chronic case of pulmonary fibrosis, Laura died with the wedding ring Enrico had given her in 1928 still on her hand.

According to her admiring granddaughter Olivia, Laura had tried to make the world a better place, looking to the future rather than dwelling on the past. The past had entailed some heartache for her. Nella and Judd, in different ways, had struggled with being the children of someone famous. This phenomenon is well documented with examples through time of how the stardom of a famous parent impacts his or her offspring. Nella and Judd were no exception. They had experienced the toll of many moves, a secret and sequestered life, and the gradual revelation that their father played a key role in developing a weapon capable of wiping out thousands of innocent lives. But perhaps most hurtful was seeing their father idolized by his students with whom he spent more time than with them. And Fermi seemed to enjoy talking physics with his followers more than he did playing a parental role with his own children. While young physicists may have been reconciled to Fermi's standard lack of praise, that trait must have been hard on Nella and Judd.

As children of a genius, both Nella and Judd felt high expectations laid on them. As adults, they were reluctant to talk about their father, even to their respective spouses or their own children. Judd harbored regrets about becoming a scientist because of obvious comparisons to his father. After earning a Ph.D. in molecular biology in Berkeley and several subsequent career moves, he settled down in England, where he once again took up molecular biology, working on Cambridge's distinguished Medical Research Council. According to his widow, Sarah, Judd was distant from both his parents.

Nella's relationship with her parents was less fraught, in part because—as she stated—less was expected of her because of her gender. In later years, Nella recounted that her father could approach her only on an intellectual level, for instance by teaching her algebra. Wisely, she had selected an entirely different field than her father's, one of which he had limited understanding or even appreciation. Her passion was for art, and she taught for thirty years at the University of Chicago's Laboratory School, the same school she and Judd had attended. She was close to her mother, undoubtedly reinforced by their proximity in Hyde Park and by Laura's frequent babysitting of Nella's two children. After her mother's death, Nella earned a doctorate in educational psychology from the University of Chicago at age fifty, a feat she was justifiably proud of. She had done what her father long ago had instructed her to do: "Earn a Ph.D., just in case." By this time, Nella had divorced, and the Ph.D. offered a sense of security, although she never seemed to overcome a certain awkwardness and lack of comfort with herself.

Throughout his dying days, Fermi was clear-headed. Laura was especially grateful for that when on November 16 he became the first recipient of a new prize instituted by the Atomic Energy Commission. Carrying an award of $25,000, the honor was to "recognize career contributions to science, technology and medicine related to nuclear energy." Fermi's letter thanking the AEC was the last he would write. Two years later the prize was renamed the Enrico Fermi Award.

Early on Sunday, November 28, 1954, Fermi had a fatal heart attack. Yet again, the Pope had been right. It was two days short of the calculations he made to Laura for his hospital bed rental. And he was two months past his fifty-third birthday. The time of his death coincided with the United Nations' formation of a committee on peaceful uses of atomic energy and with Senator Joseph McCarthy's denunciation by Congress. The world was on the verge of righting itself.

Fermi was buried in a private ceremony at Oak Woods Cemetery on Chicago's South Side. On Friday, December 3, a memorial service was held at the University of Chicago's Rockefeller Chapel. Judd had flown in from Oberlin. Even he, who assiduously avoided the many honorific events acknowledging Fermi, could not evade this one. Nella was at her mother's side, offering whatever support she could. Appropriately, the service was not far from the squash court under Stagg Field. And it was exactly one day and twelve years after the power of the atom was unleashed, heralding the birth of the atomic age.

In announcing the triumph of the pile experiment, the head of Chicago's Metallurgical Laboratory, Arthur Compton, had reported that "the Italian navigator has just landed in the new world." It was a world forever changed. And the Pope had created his transformative legacy.

AFTERWORD

I n this book, Bettina and I have sought to describe a remarkable man who shaped history, along with a turbulent history that shaped the man. Enrico Fermi was both a creator and a product of his times. A great deal is known about his astonishing intellect, but less is shared about his human side. We have combined both aspects of him, hoping to present a fuller portrayal of Fermi within the context of a mutable world.

Writing this book has been a personal journey for Bettina and me, most obviously because we are coauthors, but also on another level, since Bettina and I had each met Fermi. I met him only once and barely remember the details. Bettina met him a few times and barely remembers the details. My excuse is that I was not yet two years old. Bettina's is that she was a teenager, oblivious to fame.

For both of us, under very different circumstances, Fermi was a household name. My family had known and admired Fermi in Italy; Bettina's family knew and admired Fermi in the Atomic City of Los Alamos.

My family and the Fermis had escaped Italy in 1939. In both cases their flight had been prompted by the passage in 1938 of Italy's anti-Semitic laws. Fermi was not Jewish, but his wife, Laura, was; my father was Jewish, although my mother was not.

Like many European émigrés, they sought a permanent position in the United States. My father, Angelo Segrè, had been a professor of ancient history in Italy and landed a modest appointment at Columbia University. Fermi, a Nobel Prize winner, joined Columbia's physics department as a celebrated member.

The families had met through Emilio Segrè, the younger brother of my father. My uncle was Fermi's first student in Rome and the two physicists became lifelong friends. Emilio also had fled Italy in 1938, taking a position at the University of California in Berkeley. Emilio and Enrico had many parallels in their lives and an abiding trust in one another. They were both part of the Los Alamos Scientific Laboratory team whose work led to far-reaching consequences.

After the war my family returned to Italy. Growing up there, I came to appreciate what a source of pride Fermi was for his homeland. Fermi's 1938 Nobel Prize constituted a tangible symbol of Italy's scientific renaissance after a long drought. Like many of my Italian contemporaries, I wanted to be a physicist. And I became one.

Physics is what led to me meeting my wife, Bettina Hoerlin. Her parents, too, had fled a fascist regime. Herman Hoerlin and his Jewish wife, Kate, escaped Germany in 1939 to come to America. Hoerlin, a physicist who had been employed in industry, joined the postwar Los Alamos laboratory as a group leader. A physics colleague of mine knew the Hoerlins from Los Alamos summers. Luckily, he introduced me to Bettina.

During the summer of 1953, Fermi's last in New Mexico, he—along with other emigrant European scientists and their families—typically went on weekend hikes in the high mountains. Bettina's father was a natural leader for these excursions, given that his bona fides extended beyond physics to mountaineering; he had made first ascents in the Alps, Andes, and Himalayas. Among his fellow hikers in New Mexico, Hoerlin saved his greatest compliment for the Italian Nobel laureate, declaring in his German accent, "I never had to vorry [sic] about Fermi."

Bettina and I have attempted to approach Fermi in terms of person and place. Interviews with family members and with people who knew him or his children provided rich fodder. We followed Fermi's (physical) footsteps, first in Italy and then in America.

We traveled to the ancient city of Pisa, where Fermi entered the world of science at the prestigious Scuola Normale. We strolled around the city famous for another scientist, Galileo, who inspired the seventeen-year-old. And we held Fermi's meticulously recorded notebooks that hinted at his future thoroughness.

In Rome, Bettina and I admired the Campo dei Fiori, where fourteen-year-old Enrico bought his first physics book, and crisscrossed Rome to Via Panisperna, spotting the villa that once housed the university's Institute of Physics. It was easy to imagine the collegiality he experienced with his equally young peers, known as the Boys. The instruments used in their groundbreaking radioactive research are now lovingly displayed in a museum at the current university site.

When Bettina and I traveled to the University of Chicago to delve into the Fermi archives, it was another kind of experience. We re-created the bike ride made by Fermi from his comfortable home to work every day. The football stadium at Stagg Field, where the first nuclear pile went critical, has been torn down and in its place stands the university library and a large monument called Nuclear Energy. The silence of this symbolic amalgam of a mushroom cloud and a human skull speaks to a sense of foreboding and doom. Circling its eight-foot circumference, we understood the aim of its sculptor—Henry Moore—to create a cathedral-like experience, a tribute to the human spirit.

Reportedly, Laura Fermi was not fond of it, because it was "confusing atomic energy, which began with the Chicago Pile-1 experiment, with the birth of atomic weapons, which began with Trinity." It is difficult to separate those two events, one that built upon the other, ushering in the atomic age. In each, Fermi played a key role, although more so with the pile.

In 1944, Fermi and his family went from Chicago to Los Alamos. The town is much changed now. Unchanged is Trinity, a few hours south, where the first nuclear bomb exploded. A monument also marks this history. An obelisk sits in the epicenter of ground zero, amid bits of trinitite—a residue mineral formed by the heat of the blast. The obelisk, over twelve feet tall, happens to approximate the height of the

Nuclear Energy monument. In contrast to the latter's sleek surface of rounded bronze, the obelisk is geometric, rough-sided, and constructed from local lava rock. The site is rarely open to the public, eerily quiet, fenced off. Adding to its surreal aura, a species of oryx roam the surrounding bleak desert.

Major national—and international—disputes regarding immigration, secrecy, and nuclear weaponry were in the news almost every day during the period when Bettina and I were writing this book. They were also the issues of Fermi's day, when the atomic age was born. We seem not to have learned how to deal with them.

Today almost ten countries have atomic bombs; several others have the know-how but have either chosen not to make them or been deterred from doing so. So far, the threat of mutual annihilation has prevented use of nuclear bombs by one nation against another. Let us hope it continues that way. As articulated by Fermi, the future hinges on whether humankind will make "good use of the powers . . . acquired over nature."

April 2016

NOTES

To distinguish works by the same author, the publication date of the reference is given, for example, Maltese (2003) and Maltese (2010). The publication year is otherwise omitted.

ABBREVIATIONS

ARCF Enrico Fermi Collection, Regenstein Library, University of Chicago.

EFP 1 Enrico Fermi. 1962. *Collected Papers, Volume 1,* edited by Emilio Segrè (editor in chief), Edoardo Amaldi, Herbert Anderson, Enrico Persico, Franco Rasetti, Cyril Smith, and Albert Wattenberg. University of Chicago Press.

EFP 2 Enrico Fermi. 1965. *Collected Papers, Volume 2,* edited by Emilio Segrè (editor in chief), Edoardo Amaldi, Herbert Anderson, Enrico Persico, Franco Rasetti, Cyril Smith, and Albert Wattenberg. University of Chicago Press.

ES Emilio Segrè. 1970. *Enrico Fermi, Physicist.* University of Chicago Press.

LF Laura Fermi. 1954. *Atoms in the Family: My Life with Enrico Fermi.* University of Chicago Press.

OHI Oral History Interviews, American Institute of Physics.

PROLOGUE: TRINITY

2   "Suddenly there . . . than the eye": Rigden, p. 156.

2   "For a moment . . . not possible": ES, p. 147.

2   "I am become . . . destroyer of worlds": Rhodes (1986), p. 676. Quoted in Len Giovanitti and Fred Freed, *The Decision to Drop the Bomb*. New York: Coward McCann, 1965.

PART 1: ITALY, BEGINNINGS

The history of the Fermi family and Enrico Fermi's subsequent life has been told admirably from a physicist's point of view in Emilio Segrè's *Enrico Fermi, Physicist*. For a comprehensive view of Italian history, see Denis Mack Smith's very readable *Modern Italy*.

2: THE LITTLE MATCH (IL PICCOLO FIAMMIFERO)

13   "small . . . frail-looking": LF (1954), p. 15.

13   "to stop . . . not tolerated": Ibid.

3: LEANING IN: PHYSICS AND PISA

21   "During the first . . . my self-control": Letter from Fermi to Persico, December 12, 1918. ARCF, Box 9, folder 2.

4: STUDENT DAYS

24   "I have met . . . he understands everything": Interview with Rasetti, April 8, 1963, OHI.

26   "I have a lot to . . . piece of filth": Fermi to Persico, March 16, 1922. ARCF, Box 9, folder 2.

27   "Spaghetti and Levi-Civita": Jackson, p. 11.

5: THE YOUNG PROTÉGÉ

29   "I became a Senator . . . surrounded by apparatus": ES, p. 30, quoting Corbino in *Conferenze e Discorsi* (1937), p. 167.

31   "were very conscious . . . to stress the point": Interview with Amaldi, April 8, 1963, OHI.

32   "I could never . . . morning to be a theoretical physicist": Commentary by Harold Agnew, January 6, 1955. ARCF, Box 7, folder 2.

6: THE SUMMER OF 1924

36 "I have finished . . . for my funeral": LF (1961), p. 229.

36 "Italy wants . . . forty-eight hours": LF (1961), p. 245, and Smith (1970), p. 332.

37 "In the first few years . . . which happened in 1924": Rasetti (1982), p. 24.

38 "So many young men . . . than Fascism": LF (1954), p. 30.

38 "A hope . . . under Mussolini": Ibid.

38 "Better a day as a lion than a hundred years as a sheep": Italian proverb quoted by Mussolini on June 20, 1926.

38 "promotion and . . . throughout the world": Kevles (1995), p. 83.

39 "really very nice . . . in a ghetto store for used clothing": Letter from Fermi to Persico, October 23, 1924. ARCF, Box 9, folder 2.

40 "very nice person . . . (pity that he's not a beautiful girl)": Ibid.

40 "the characteristics . . . missing hypotheses": Persico introduction to EFP1, p. 142.

7: FLORENCE

43 "an act of desperation": Hermann (1971), p. 74.

45 "the first instance . . . of radiofrequency fields": EFP1, p. 159.

45 "given the . . . very pleasing": Letter from Fermi to Persico, October 15, 1925. ARCF, Box 9, folder 2.

8: QUANTUM LEAPS

49 "There was a moment . . . rise and was happy": Van der Waerden (1967), p. 25.

50 "My impression . . . zoology of spectroscopic terms": Letter from Fermi to Persico, September 29, 1925. ARCF, Box 9, folder 2.

50 "Now I'm trying to . . . so far I don't understand it": Interview with Rasetti, April 8, 1963, OHI.

50 "This was . . . perfect, admirable": Born (1978), p. 226.

51 "I was absolutely . . . if not to say repelled": Schrödinger (1926), p. 755.

51 "understood it and then . . . few people around him": Interview with Rasetti, April 8, 1963, OHI.

53 "Italy should have . . . scientific achievement": Segrè (1993), p. 46.

53 "I have serious . . . and trotting out Mussolini": Eckert (2013), p. 301.

54 "Everybody began . . . through Germany": Interview with Rasetti, April 8, 1963, OHI.

9: ENRICO AND LAURA

57   "We are going . . . as my father says": LF (1954), p. 6.

58   "He shook . . . and amused": Ibid., p. 3.

60   "dispassionate detachment . . . or strange plants": Ibid., p. 34.

60   "country stock": Ibid., p. 52.

60   "When I first met her . . . I had ever met": Libby (1979), p. 28.

60   "Enrico caught his breath . . . teen-age Laura was": Ibid., pp. 28–29.

61   "If she wanted . . . make it for herself": ES, p. 33.

62   "Congratulations, Mrs. Fermi": LF (1954), p. 57.

PART 2: PASSAGES

For more background on Italian Fascism, consult Laura Fermi's interesting book *Mussolini*. Nuclear physics history during the period between the discoveries of the neutron and of fission is covered extensively in Edoardo Amaldi's "From the Discovery of the Neutron to the Discovery of Nuclear Fission" in *Physics Reports*.

10: THE BOYS OF VIA PANISPERNA

66   "I believe . . . poetic feelings about it": ES, p. 53.

66   "by tailing Rasetti . . . what was going on": Segrè (1993), p. 45.

11: THE ROYAL ACADEMY

70   "What practical . . . matter's intimate structure?": EFP1, p. 371.

71   "that the work . . . purely abstract ideas": Ibid., p. 377.

71   "Enrico felt . . . be prepared for emergencies": LF (1954), p. 60.

71   "I was to learn . . . about physics": Ibid., p. 58

72   "was mediocre prose . . . returns for many years": Ibid., p. 62.

72   "I am the driver . . . let me in": Orear, p. 78.

75   "I had once to take . . . I always regretted it": Wick, quoted in Jacob, p. 11.

12: CROSSING THE ATLANTIC

78   "Almost my . . . by Fermi": Feynman (2005), p. 377.

79   "His ability to . . . literature is infallible": Schweber (2012), p. 193.

13: BOMBARDING THE NUCLEUS

81   "The study . . . for the physics of tomorrow": ES, pp. 65–67.

81   "we should expect . . . of nuclear phenomena": EFP1, p. 361.

82   "These numbers . . . energetic chemical bonds": Ibid., p. 33.

83   "the depth and universality of Italian thought": Cordella (2001), p. 202.

85   "I don't believe . . . entirely out of character": Chadwick (1964), p. 161.

85   "For the neutron . . . something else": Segrè (1980), p. 184.

85   "They haven't . . . heavy neutral particle": Maltese (2010), p. 197.

86   "I forbid you to mention . . . go around discrediting me": Ibid., p. 199.

14: DECAY

88   "Any physicist . . . Just the facts!": Goldberger, quoted in Cronin (2004), p. 158.

90   "it contained . . . too remote from physical reality": EFP1, p. 540.

91   "One can . . . inside the nucleus": Bethe (1934), p. 532.

91   "There is no . . . the neutrino": Ibid.

15: THE NEUTRON COMES TO ROME

93   "The problem . . . intellectual torpor": Letter from Fermi to Segrè, September 30, 1932. ARCF, Box 11, folder 13.

95   "To Professor . . . slow neutrons": Fermi 1939 Nobel Prize in Physics citation.

97   "I congratulate . . . sphere of theoretical physics": EFP1, p. 641.

97   "Now we will all have to learn Italian": Holton, quoting Rabi in Bernardini (2001), p. 63.

98   "The Pope is upstairs": LF, p. 89.

99   "The investigation . . . of this new element is certain": Maltese (2010), p. 269.

99   "The public . . . to study the general phenomenon": LF (1954), p. 92.

16: THE RISE AND FALL OF THE BOYS

103   "What a stupid . . . known to foresee it": Cordella, p. 244.

103   "intuito fenomenale": Maltese (2010), p. 266.

108   "worked with . . . part in the Spanish Civil War": EFP1, p. 811.

108   "The paper contains. . . . in succeeding years": Ibid., pp. 810–11.

109   "dare to take . . . bewilderment and misgivings": LF (1954), p. 105.

17: TRANSITIONS

110 "Fermi had . . . reticence": Segrè, quoted in Cronin (2004), p. 25.

113 "hopeless to think . . . these researches on an adequate basis": Maltese (2010), p. 308.

113 "that are . . . one in Denmark": Ibid., p. 309.

115 "in one single unshakable determination": LF (1961), p. 352.

18: STOCKHOLM CALLS

116 "I had . . . firmly rooted in Rome": LF (1954), p. 113.

119 "for reasons easily understood": Maltese (2003), p. 14.

119 "Making use of my . . . because of the racial laws": Ibid.

121 "a great honor . . . Duce . . . so I can take . . . in scientific circles of those countries": Letter from Fermi to Oswaldo Sebastiani, December 3, 1938. ACRF, Box 3, folder 12.

122 "I hope I'll see you soon": LF (1954), p. 129.

122 "one or more elements . . . of Ausonium and Hesperium respectively": Nobel Prize in Physics lecture (by Fermi), 1938.

123 "I thank . . . just like my preceding five visits . . . without special warmth": Letter from Federzoni to Mussolini, January 5, 1939. ACRF, Box 3, folder 13.

PART 3: HELLO, AMERICA

For a general background to part 3, we can recommend two very good books by physicists and friends of Fermi: Arthur Compton's *Atomic Quest* and Leona Marshall Libby's *The Uranium People*.

19: FISSION

128 "She taught me . . . the beryllium foil and so on": Rasetti (1982), p. 13.

128 "a worried, tired . . . in common": LF (1954), p. 156.

129 "Your radium . . . leads to barium!": Meitner, quoted in Sime (1996), p. 235.

130 "a drop of fluid . . . surface tension": Bohr, vol. 9, p. 47.

131 "Oh what idiots . . . a paper about it?": Frisch, p. 116.

133 "probably a scientist not discovering fission": Allison, p. 129.

133 "for his . . . of heavy nuclei": 1944 Nobel Prize in Chemistry citation.

20: NEWS TRAVELS

135 "gracious way of life": Libby, p. 22.

136 "Enrico had often . . . raised objections": LF (1954), p. 116.

136 "We have . . . of the Fermi family": Ibid., p. 139.

138 "Bohr has gone crazy . . . uranium nucleus splits": Blumberg (1976), p. 46.

139 "Young man . . . overwhelmed by all that": Interview with Anderson, January 13, 1981, OHI.

139 "Simply because . . . to explain later": Pais (1991), p. 45.

139 "Nothing like . . . else in the world": Interview with Anderson, January 13, 1981, OHI.

139 "So Fermi . . . and work everything out": Ibid.

140 "mumbled and . . . implications": As quoted in Rhodes (1986), p. 271.

141 "LINEAR AMPLIFIER . . . INFORMATION POSTED": Bohr, vol. 9, p. 551.

141 "I need not . . . you most heartily": Ibid., p. 563.

141 "It was suggested . . . division of bacteria": Ibid., p. 559.

142 "After this . . . some days before": EFP1, p. 3.

142 "The phenomenon . . . of the order of 200 Mev": EFP2, p. 2.

142 "merit was . . . interest of all physicists": Bohr, vol. 9, p. 554.

143 "No great human concern left Bohr indifferent": French, p. 226.

21: CHAIN REACTION

145 "A little bomb like that and it would all disappear": Kevles (1979), p. 324.

145 "some German chemists . . . bombarded with neutrons," "a lot of theoretical reasons . . . couldn't really happen," and "When I invited . . . the right conclusions": Interview with Alvarez, February 15, 1967, OHI.

148 "I have often . . . rearm and go to war": Weart (1980), p. 19.

149 "Ten percent is . . . that we may die from it": Ibid., p. 54.

149 "Fermi thought . . . all the necessary precautions": Ibid.

150 "can never be done unless you turn the United States into one huge factory": Rhodes (1986), p. 294.

22: THE RACE BEGINS

151 "There's a wop outside": Rhodes, p. 295.

154 "between the end of . . . in natural uranium": Weart, p. 115.

154 "Because cosmic rays . . . free and everywhere": Rasetti (1982), p. 47.

155 "This is a big and free country . . . and begin again": Fermi, quoted in Heisenberg (1971), p. 170.

155 "Where to?" . . . "Somewhere . . . the world is large": LF (1954), p. 31.

155 "found it . . . behind such a project": Heisenberg (1971), p. 193.

155 "Don't you think it . . . he does not want to see": Ibid., p. 170.

156　*"Daran habe ich gar nicht gedacht"* ("I hadn't thought of that at all"): Clark, pp. 669ff.

156　"In the course of . . . bombs of a new type may thus be constructed": As cited in Lanouette, p. 205.

157　"Alex, what you are after is to see that the Nazis don't blow us up." Sachs replied, "Precisely": Lanouette, p. 210.

158　"it was very interesting . . . that budget could be cut": Weart, p. 85.

23: NEW AMERICANS

159　"Among adult immigrants . . . toward Americanization": ES, p. 104.

160　"more freedom . . . You can't make me wash my hands. This is a free country": Laura Fermi, quoted in Orear, p. 149.

160　"on the Palisades . . . the basement": LF (1954), p. 145.

160　"his peasant blood was not aroused": Ibid.

161　"We have noticed . . . when he'll return": Rome police to Ministry of Foreign Affairs, March 7, 1939. ACRF, Folder 16, box 4.

161　"Prof. Fermi . . . enemies of our Regime": Maltese (2003), p. 76.

162　"You wouldn't put . . . or would you?": Weart, p. 143.

163　"always take place . . . if the size of the pile is large enough": EFP1, p. 225.

165　"to dark corridors . . . experiment": Ibid., p. 269.

166　"started looking . . . what was happening": Ibid.

24: THE SLEEPING GIANT

168　"The government's responsible . . . war program": Compton (1956), p. 49.

170　"We estimated . . . after all, at least in principle!": Peierls, pp. 154–55.

171　"inarticulate and unimpressive": Oliphant, p. 17.

171　"I thought we were . . . for submarines": Davis, p. 112.

172　"No one can . . . well as Enrico Fermi": Compton, p. 11.

172　"simply and directly . . . chain-reacting sphere": Ibid., p. 54.

172　"almost with tears . . . knew them so well": Ibid., p. 55.

172　"felt like swimming in syrup": Szanton, p. 205.

172　"Even by . . . so were sadly mistaken": Weart, p. 147.

25: CHICAGO BOUND

176　"speaking with equal ease . . . American in both her habits and way of thinking": Fermi to Amaldi, April 5, 1941, in Battimelli (1997), p. 130.

177 "he wished Hitler and Mussolini would win the war": LF (1954), p. 173.

178 "His associates like him . . . is not recommended": Lanouette, p. 223.

179 "was filthy with dirt . . . swindled by a slick sales talk": EFP1, p. 206.

179 "doing physics by the telephone": ES, p. 121.

180 "Perhaps the . . . of taxes and death": Libby, p. 1.

180 "gaiety and informality . . . usually calm and mildly amused": Leona Woods Marshall, quoted in EFP1, p. 328.

181 "I had been . . . I had difficulty just walking back": Cronin, p. 185.

181 "Laura fed us supper after supper": Libby, p. 7.

182 "the commitment of . . . quite a mess of machinery": Rhodes (1986), p. 407.

182 "If you do the job right, it will win the war": Groves, p. 4.

182 "I felt there . . . instead of as a promoted colonel": Ibid.

26: CRITICAL PILE (CP-1)

184 "I have . . . strongly affected me": Compton, p. 10.

185 "lack of interest in the physical world": Libby, p. 24.

185 "fresh, clear and convincing . . . for the job at hand": Herbert Anderson, quoted in EFP1, p. 216.

185 "run quick-like . . . many miles away": EFP1, pp. 169–74.

186 "If people could . . . they'd know we are": "The First Pile," in http://www.atomicarchive.com, p. 1.

187 "These tough . . . negative commitment to work" and "The graphite machining . . . Back of the Yards kids": Wattenberg (1982), p. 22.

187 "It was a privilege . . . him in those days": Herbert Anderson, quoted in EFP1, p. 216.

191 "Will the reaction . . . self-sustaining . . . be thermally stable . . . be controllable?" EFP1, p. 263.

27: THE DAY THE ATOMIC AGE WAS BORN

195 "I'm hungry. Let's go to lunch": Wattenberg (1982), p. 31.

196 "Again and again . . . 'The pile has gone critical'": Anderson (1974), p. 42.

196 "alert, in as full . . . of the work": Compton, p. 143.

197 "His eyes were aglow . . . the wheels of industry": Ibid., p. 144.

197 "'Jim,' I said, 'you'll be interested to know that the Italian navigator . . . Everyone landed safe and happy'": Ibid.

198 "What's going on, Doctor, something happen in there?" Laurence, p. 71.

PART 4: THE ATOMIC CITY

For the history of the atom bomb, there is no substitute for the magisterial Richard Rhodes account, *The Making of the Atomic Bomb*. For a magnificent set of photographs and stories about the Manhattan Project, see Rachel Fermi and Esther Samra's *Picturing the Bomb*.

28: THE MANHATTAN PROJECT: A THREE-LEGGED STOOL

201   "The chain reacting . . . then in a satisfactory way": EFP1, p. 270.

201   "The story of . . . of human knowledge" and "It was the good fortune . . . of the atomic age": Edward R. Murrow, *CBS Evening News*, December 2, 1954. ACRF, Box 7, folder 1.

202   "He sank a Japanese admiral": Libby, p. 129.

202   "anything was possible for Enrico": Ibid.

203   "What thrilled . . . experimental tool": Herbert Anderson, quoted in EFP2, p. 308.

203   "To operate a pile is just as easy . . . on a straight road": EFP2, p. 548.

204   "the work he enjoyed most": Herbert Anderson, quoted in EFP2, p. 352.

208   "My two great loves . . . can't be combined": Rhodes (1986), p. 451.

208   "If you go on . . . be a usable site": Bird, p. 206.

29: SIGNOR FERMI BECOMES MISTER FARMER

210   "absolutely unscrupulous": Bird, p. 213.

210   "We were faced . . . and living conditions": Rhodes (1986), p. 450.

210   "The projected laboratory . . . a real, live, breathing Enrico Fermi": Robert Wilson, p. 41.

211   "The object of the . . . nuclear fission": Serber (1992), p. xi.

211   "I believe your people . . . want to make a bomb": Rhodes (1986), p. 468.

212   "One evening . . . he would wait for him outside": LF (1954), p. 212.

212   "By the time Baudino . . . straight-faced about this debt": Libby, p. 162.

213   "My work clothes . . . concealed the bulge," "because Zinn would . . . of the reactor building," and "When he told me . . . practice midwifery": Libby, p. 164.

213   "absolutely and entirely renounce and abjure all allegiance and fidelity to any foreign prince, potentate, state, or sovereignty, of whom or which I have heretofore been a subject or citizen": http://usgovinfo.about.com/od/immigrationnaturalizatio/a/oathofcitizen.htm.

214 "I haven't the faintest idea": LF (1954), p. 201.

216 "While the Pope is not . . . and there it stops": Ibid., p. 215.

218 "sent a card from Ferrara . . . competent sources in Washington": Letter from Sandro Motel (married to Laura's sister) to Fermi, November 15, 1944. ARCF, Box 10, folder 11.

218 "As you can imagine . . . worse than knowing him dead": Battimelli (1997), pp. 148–49.

30: GÖTTERDÄMMERUNG

220 "How can you . . . your whole life?": Bernstein (2001), p. 40.

220 "Though a brilliant . . . about numbers": Ibid., p. 36.

223 "Enrico Fermi was our anchor man on such occasions": Compton, p. 191.

223 "since this had . . . through by Fermi": Herbert Anderson, quoted in EFP1, p. 428.

224 "It is quite certain . . . will arrive next week": Ulam (1970), p. 162.

224 "I was ready to sell my soul": Robert Wilson, quoted in Orear, p. 108.

225 "declined to take charge . . . calculations on the implosion": Rhodes (1986), p. 543.

31: THE HILL

226 "totally absorbed, taking little notice of his family": Libby, p. 27.

226 "It was typical . . . was going on around him": LF (1954), p. 82.

227 "The rickety houses looked like . . . slum area; everywhere . . . garbage cans were overflowing": Marshak, quoted in Jane Wilson (1997), p. 9.

227 "Through the . . . an old master": LF (1954), p. 207.

227 "We managed . . . happy lives": Jane Wilson (1997), p. 43.

227 "Lines were drawn . . . in the Laboratory": Marshak, quoted in Jane Wilson, p. 9.

228 "danced with . . . instead of his feet": Brode, quoted in Rhodes (1986), p. 564.

228 "Saturday nights . . . dances and parties": J. Wilson, p. 112.

230 "You see . . . learn their tricks" and "I see, matching wits!": Orear, p. 99.

230 "he was . . . unpretentious way could be so dominant": Frisch, p. 167.

230 "never appeared . . . because he was so organized": Ibid.

230 "I told Fermi . . . was quite a lesson for me": Feynman, p. 132.

231 "Bohr immediately . . . task of peace that lay ahead": Weisskopf, p. 144.

232 "Bohr talked . . . priest celebrating mass": Ulam, p. 167.

232   "we hire girls . . . they're cheaper": Howes, p. 99.
232   "depressed, quarrelsome and gossipy" and "keep the women busy . . .
       proof of their usefulness": Segrè, p. 190.

32: "NO ACCEPTABLE ALTERNATIVE"

234   "we all trusted him and saw him frequently": LF (1954), p. 210.
234   "no more than . . . on atoms was involved": Francis Sill Wickware,
       "The Manhattan Project," *Life*, vol. 19 (August 20, 1945), p. 26.
235   "the development of new explosive . . . destructive power": Truman, p. 10.
235   "the whole purpose . . . ceased to be": Rotblat, quoted in Kelly, p. 280.
237   "not bothered . . . would be used": Rotblat (1985), 18.
237   "without prior warning": Rhodes (1986), p. 651.
237   "a man of great wisdom . . . to shorten this war" : Ibid., pp. 625–26.
238   "sensitized to . . . chemical weapons program": Von Hippel, p. 41.
239   "We believe that . . . control of such weapons": Kelly, p. 288.
239   "we can propose . . . direct military use": Ibid., p. 291.
239   "With regard to . . . by the advent of atomic power": Ibid., p. 290.
241   "after all . . . atomic explosion was not possible": Groves, p. 297.
242   "so sleepy he went to . . . his strong aversion to being driven": LF
       (1954), p. 238.
242   "governing factor . . . previously damaged by air raids": Groves, pp. 267ff.

33: AFTERSHOCK

244   "I saw a bright blast . . . blew away all at once": Michiko Kodoma on the
       bombing of Hiroshima, August 10, 2015, www.rifuture.org.
244   "The general impression is one of deadness . . . it is not so here": Navy
       Captain William C. Bryson, *Bulletin of the Atomic Scientists*, Decem-
       ber 1982, p. 35.
244   "The first the general public . . . identified with death and destruc-
       tion": Rotblat and Ikeda, p. 32.
244   "I don't believe . . . whole thing": Bernstein (2001), p. 116.
245   "it was dreadful of the Americans . . . it's madness on their part":
       Weiszäcker, quoted in Bernstein (2001), p. 117.
245   "One can't say . . . 'ending the war'": Bernstein (2001), p. 117.
245   "I am thankful . . . to drop the uranium bomb": Ibid., p. 125.
246   "the least abhorrent choice . . . a million casualties to American forces
       alone": Article by Henry Stimson, *Harper's Magazine*, February 1947,
       quoted in Kelly, p. 409.

246   "When the [atomic] bombs . . . grow up to adulthood after all": Folsom, p. 310.

246   "Well at last you know . . . even on our enemy": Kelly, p. 342.

246   "Children . . . with lids and spoons": LF (1954), p. 240.

247   "It seemed rather ghoulish . . . even if they were 'enemies'": Frisch, p. 176.

247   "romantic": Robert Wilson, p. 41.

247   "nervous wreck": Bird, p. 317.

247   "among the praising . . . heard from several directions": LF (1954), p. 240.

247   "But above all, there were the moral questions . . . there is no simple answer": Laura Fermi, quoted in Badash, p. 89.

247   "All however are perplexed . . . Who alone can judge you morally": Maria Fermi Sacchetti, quoted in LF (1954), p. 245.

248   "It seems . . . face severe judgment by history": Maltese (2003), p. 172.

248   "There are few decisions . . . less cause for regret": Ibid.

249   "the needs of national security" and "subject to . . . Espionage Act": Smyth Report, p. v.

249   "Don't be afraid of becoming . . . blow up too": LF (1954), p. 237.

250   "should not end . . . remained a military secret": Kelly, p. 285.

250   "directing and encouraging the use of atomic energy . . . peaceful and humanitarian ends": President Truman, Special Message to Congress on Atomic Energy, October 3, 1945.

251   "any change . . . into this category": Libby, p. 256.

251   "were duped . . . silent about the army's bill": Lanouette, p. 287.

251   "For me it was . . . to be relied upon": Bird, p. 327.

252   "The struggles of our times . . . and he is no fighter": Weart, p. 146.

252   "Whatever Nature . . . better than knowledge": Enrico Fermi, quoted in LF (1954), p. 244.

252   "a labor of considerable scientific interest" and "having contributed to truncating a war . . . a certain satisfaction": Letter of Fermi to Amaldi, August 28, 1945, in Battimelli (1997), pp. 158–60.

34: GOODBYE, MR. FARMER

253   "It had been one . . . periods of our lives": LF (1954), p. 246.

255   "if Enrico . . . would profit by the exchange": Compton, p. 203.

257   "At Los Alamos, where I had been working . . . probably help them too": Wilson, quoted in U. Amaldi, p. 225.

257   "we all felt that . . . we had done our duty": Bethe (1982), p. 45.

PART 5: HOME

As the reader has seen, Laura Fermi's delightful and informative *Atoms in the Family: My Life with Enrico Fermi* has been a great source of material throughout the book. The background to the buildup of weapons and related issues is brilliantly recounted in Kai Bird and Martin Sherwin's *American Prometheus: The Triumph and Tragedy of J. Robert Oppenheimer.*

35: PHYSICIST WITH A CAPITAL *F*

262   *The Quick and the Dead*: *The Quick and the Dead*, WMAQ radio, newspaper column (undated, unmarked) by Bill Irvin, Radio-Television News. ARCF, Box 1, folder 1.

262   "the most important . . . the birth of Jesus Christ": Rhodes (1995), p. 279.

262   "Your scientific skill . . . as the agent of the American people": Letter from Major General Groves to Fermi, September 28, 1945. ARCF, Box 4, folder 7.

264   "an ambassador of mathematical beauty": Farmelo, p. 435.

264   "Fermi was the Pied Piper who brought them": Cronin, p. 153.

265   "What would you like? . . . You name it": Cronin, p. 169.

265   "as pleased . . . to upset his routine": LF (1954), p. 258.

267   "The history of . . . an exception to this rule": Cronin, p. 142.

267   "What is less . . . that he acquires over nature": Ibid.

36: THE FERMI METHOD

268   "my first physics . . . right decision!": Cronin, p. 237.

269   "If I am to be . . . because of your training": Ibid., p. 187.

269   "Physics is to be . . . layer by layer": EFP1, p. 239.

270   "They were . . . marks on my memory": Cronin, p. 197.

270   "perfectly normal": LF (1954), p. 227.

270   "with a minimum of complication and sophistication": Victor Weisskopf, quoted in Bernardini (2001), p. 5.

271   "How thick can the . . . before it falls off?": Cronin, p. 179.

271   "What is the number of sheep in Nevada?": Libby, p. 16.

272   "How many piano . . . in the city of Chicago?": Cronin, p. 179.

272   "we could solve any problem": Ibid.

272   "Fermi had an inside track to God:" Orear, p. 29.

272 "certainly not the person to go to" ... "not cold, but not warm either": Goldberger, quoted in Cronin, p. 155.

272 "a musician ... long practice and study": EFP2, p. 923.

272 "I believe ... in the last five years": Letter to Fermi from George Kistia-kowsky, Harvard Department of Chemistry, September 29, 1953. ARCF, Box 10, folder 10.

273 "we must be prepared for a long pull": EFP2, p. 834.

273 "There are two ways ... You have neither": Dyson, p. 297.

273 "I remember my ... stuck in a blind alley": Ibid.

274 "Enrico, they are already ... Hungarians": Marx, p. 225.

274 "What Edward is trying to say is": Cronin, p. 152.

37: THE SUPER

276 "are much more ... with senators and congressmen": Maltese (2003), p. 230.

276 "No man could ... nor more loved for himself": Commentary by Norris Bradbury. ARCF, Box 7, folder 2.

277 "We believe that the safety of this nation ... making future wars impossible": Rhodes (1986), pp. 751–52.

279 "The fact that no ... an evil thing in any light" and "to tell the American public ... such a weapon": Rhodes (1995), p. 402.

279 "could have voted ... the hydrogen bomb worked": Libby, p. 15.

279 "flared up ... shaking and speechless": Ibid.

280 "some very serious inconveniences ... proceeding very slowly": Maltese (2003), p. 230.

280 "We used to talk and talk ... it was the situation": Interview with Amaldi, April 10, 1969, OHI.

281 "blow them off the face of the earth before they do that to us": Rhodes (1995), p. 152.

281 "he was ... to have been very incomplete": Bethe (1982), p. 47.

281 "the big three": Ford, p. 152.

281 "It is ... your technical success": Rhodes (1986), p. 476.

282 "once basic knowledge ... around the sun": EFP 2, p. 556.

282 "It is impossible ... My utterly uninformed ... between two and five years": Letter from Fermi to W. Borden, Executive Director, Joint Committee on Atomic Energy, November 26, 1952. ARCF, Box 9, folder 17.

38: CIRCLING BACK

283 "Mrs. Ferrall has noticed": Letters dated May 16, 1943, May 17, 1944, and June 17, 1946. ARCF, Box 9, folder 6.

284 "forces that . . . dangerously strong," "but my knowledge . . . hear my views," and "I believe . . . democratic form of government": Letter from Fermi to the Honorable James F. Byrnes, secretary of state, October 16, 1945. ARCF Box 9, folder 17.

287 "demonstrating his . . . given its first impulse": EFP2, p. 684.

288 "What a pity that . . . we all have to be on Oppenheimer's side": Telegdi, *Physics Today*, June 2002, pp. 38–42.

289 "born with a golden spoon in his mouth": Telegdi, quoted in Orear, p.89.

289 "when I knew him . . . Acheson by his first name": Orear, p. 113.

289 "Fermi, on the other . . . was Fermi!" Ibid.

289 "defrock him in his own church": USAEC, p. 710.

289 "I would like to see . . . and therefore trust more" and "it would be wiser not to grant clearance": Bird, p. 534.

290 "In England they would have knighted him": Bethe (1968), p. 391.

290 "public-policy priesthood": Bird, p. 549.

39: LAST GIFT TO ITALY (*ULTIMO REGALO ALL'ITALIA*)

292 "smiled archly . . . how angels felt" and "began singing . . . 'seen the glory'": Glauber, *Physics Today*, June 2002, pp. 44–46.

293 "would constitute . . . inestimable way": Maltese (2003), p. 426.

293 *"ultimo regalo all'Italia"*: Ibid.

295 "I found . . . for him the place of song": Ibid., p. 427.

295 "It wasn't that . . . ability to express them": Orear, p. 129.

40: FAREWELL TO THE NAVIGATOR

298 "Tell me, Chandra, when . . . an elephant?": Cronin, p. 232.

298 "Gell-Mann and I . . . look into his face": EFP1, p. 674.

298 "being fed artificially . . . discussed it with Socratic serenity": ES, 184.

298 "It pleased them and it did not harm me": Segrè (1993), p. 252.

298 "I hope the book . . . at the right moment": Ibid., p. 252.

299 "adding with . . . 'to save a soul?'": Ibid., p. 251.

299 "the best thing Teller can do . . . that people may forget": Ibid.

299 "told her to rent it only . . . need it after that": Libby, p. 22.

300 "directing the reader . . . 'Care of St. Peter'": Ibid., p. 20.

300   "Some physicists' wives . . . some more justified than others" and "But all in all, life . . . is well worth living": Orear, p. 153.

302   "Earn a Ph.D. . . . in case": Authors' interview with Rachel Fermi, March 20, 2016.

302   "recognize career contributions . . . to nuclear energy": Fermi Award.

302   "the Italian navigator has just landed in the new world": Compton, p. 144.

BIBLIOGRAPHY

Allison, Samuel. "Enrico Fermi: A Biographical Memoir." Washington, D.C.: *Proceedings of the National Academy of Sciences*, 1957.

Alperovitz, Gar. *Atomic Diplomacy*. New York: Penguin Press, 1985.

Amaldi, Edoardo. *La Vita e l'Opera di Ettore Majorana*. Roma: Accademia dei Lincei, 1966.

——. "From the Discovery of the Neutron to the Discovery of Nuclear Fission." *Physics Reports* 111, pp. 1–332. 1984.

Amaldi, Ugo. *Particle Accelerators: from Big Bang Physics to Hadron Therapy*. New York: Springer, 2012.

Badash, Lawrence, Joseph Hirschfelder, and Herbert Broida. *Reminiscences of Los Alamos 1943–45*. New York: Springer, 1980.

Battimelli, Giovanni. "Enrico Fermi: Genius and Giant of Physics." Geneva: *CERN Courier*, September 2001, pp. 26–29.

——. *L'Eredità di Fermi*. Rome: Editori Riuniti, 2003.

Battimelli, Giovanni, and Michelangelo De Maria. *Da Via Panisperna all'America*. Rome: Editori Riuniti, 1997.

Bernardini, Carlo, and Luisa Bonolis, eds. *Enrico Fermi and the Universe of Physics*. Accademia Nazionale dei Lincei, 2001.

Bernstein, Jeremy. *Hitler's Uranium Club: The Secret Recordings at Farm Hall*. New York: Copernicus Press, 2001.

———. *J. Robert Oppenheimer: Portrait of an Enigma*. Chicago: Ivan Dee, 2004.

———. *Plutonium*. Washington, D.C.: Joseph Henry Press, 2007.

Bethe, Hans. "Enrico Fermi Remembered." *Reviews of Modern Physics* 23, July 1955, pp. 263–68.

———. "J. Robert Oppenheimer." *Biographical Memoirs of Fellows of the Royal Society* 14:391, 1968.

———. "Comments on the History of the H-Bomb." *Los Alamos Science*, Fall 1982.

Bethe, Hans, and R. Peierls. "The Neutrino." *Nature*, April 7, 1934, p. 532.

Bird, Kai, and Martin J. Sherwin. *American Prometheus: The Triumph and Tragedy of J. Robert Oppenheimer*. New York: Knopf, 2005.

Blackett, Patrick. "Rutherford." *Notes and Records of the Royal Society* 27, 1972, pp. 57–72.

Blumberg, Stanley, and Gwinn Owens. *Energy and Conflict: The Life and Times of Edward Teller*. New York: G. P. Putnam's Sons, 1976.

Blumberg, Stanley, and Louis Panos. *Edward Teller*. New York: Charles Scribner and Sons, 1990.

Bohr, Niels. *Collected Works*, vol. 9. Edited by Rudolf Peierls. Amsterdam: North Holland, 1986.

Born, Max. *My Life: Recollections of a Nobel Laureate*. London: Taylor and Francis, 1978.

Brown, Laurie. "The Idea of the Neutrino." *Physics Today*, September 1978, pp. 23–28.

Brown, Laurie, Abraham Pais, and Brian Pippard, eds. *Twentieth-Century Physics*, vols. 1–3. American Institute of Physics Press, 1995.

Brown, Laurie, and John Rigden, eds. *Most of the Good Stuff: Memories of Richard Feynman*. New York: American Institute of Physics, 1993.

Caffarelli, Roberto Vergara, and Elena Volterrani. *Enrico Fermi: Immagini e Documenti*. Pisa: La Limonaia, 2001.

Casimir, Hendrik. *Haphazard Reality: Half a Century of Physics*. New York: Harper and Row, 1983.

Cassidy, David. *Uncertainty: The Life and Science of Werner Heisenberg*. New York: W. H. Freeman, 1992.

Chadwick, James. "Possible Existence of the Neutron." *Nature* 129, 1932, pp. 312–13.

———. "The Existence of the Neutron." *Proceedings of the Royal Society of London* A136, 1932, p. 392.

———. *Some Personal Notes on the Discovery of the Neutron*. Proceedings of the Tenth Annual Congress of the History of Science in Ithaca, New York. Paris: Hermann, 1964.

Close, Frank. *Half-Life: The Divided Life of Bruno Pontecorvo, Physicist or Spy.* New York: Basic Books, 2015.

Compton, Arthur. *Atomic Quest.* New York: Oxford University Press, 1956.

Cooper, Dan. *Enrico Fermi and the Revolutions of Modern Physics.* New York: Oxford University Press, 1999.

Corbino, Orso. *Conferenze e Discorsi.* Rome: Pinci, 1937.

Cordella, Francesco, with Alberto de Gregorio and Fabio Sebastiani. *Enrico Fermi: Gli Anni Italiani.* Rome: Editori Riuniti, 2001.

Crease, Robert, and Charles Mann. *The Second Creation: Makers of the Revolution in Twentieth-Century Physics.* New York: Collier Books, 1986.

Cronin, James, ed. *Fermi Remembered.* Chicago: University of Chicago Press, 2004.

Curie, Irène, and Frédéric Joliot. "The Emission of High Energy Photons from Hydrogeneous Substances with Very Penetrating Alpha Rays." *Comptes Rendus de l'Académie des Sciences* 194, p. 273, Paris, 1932.

Davis, Nuel Pharr. *Lawrence and Oppenheimer.* New York: Simon & Schuster, 1968.

Del Gamba, Valeria. *Il Ragazzo di Via Panisperna: L'Avventurosa Vita di Franco Rasetti.* Torino: Bollati Boringhieri, 2007.

Des Jardins, Julie. *The Madame Curie Complex.* New York: Feminist Press, 2010.

Dirac, Paul. "The Fundamental Equations of Quantum Mechanics." *Proceedings of the Royal Society of London* A109, 1925, pp. 642–53.

Dyson, Freeman. "A Meeting with Enrico Fermi." *Nature* 427, 2004, p. 297.

Eckert, Michael. *Arnold Sommerfeld: Science, Life and Turbulent Times.* Berlin: Springer Verlag, 2013.

Enz, Charles. *No Time to Be Brief: A Scientific Biography of Wolfgang Pauli.* New York: Oxford University Press, 2002.

Farmelo, Graham. *The Strangest Man: The Hidden Life of Paul Dirac.* New York: Basic Books, 2009.

Fermi, Enrico. *Introduzione alla Fisica Atomica.* Bologna: Zanichelli, 1928.

———. *Molecole e Cristalli.* Bologna: Zanichelli, 1934.

———. *Nuclear Physics.* Chicago: University of Chicago Press, 1949.

———. *Collected Papers*, vol. 1. Edited by Emilio Segrè. Chicago: University of Chicago Press, 1962.

———. *Collected Papers*, vol. 2. Edited by Emilio Segrè. Chicago: University of Chicago Press, 1965.

Fermi, Laura. *Atoms in the Family: My Life with Enrico Fermi*. Chicago: University of Chicago Press, 1954.

———. *Mussolini*. Chicago: University of Chicago Press, 1961.

Fermi, Olivia. *On the Neutron Trail*. http://neutrontrail.com/olivia-fermi-on-the-neutron-trail/ and http://neutrontrail.com/.

Fermi, Rachel, and Esther Samra. *Picturing the Bomb: Photographs from the Secret World of the Manhattan Project*. New York: Harry N. Abrams, 1995.

Ferroni, Fernando. *Edoardo Amaldi in Science and Society*. Bologna: Società Italiana di Fisica, 2010.

Feynman, Richard. *Perfectly Reasonable Deviations*. New York: Perseus, 2005.

Folsom, Burton W., and Anita Folsom. *FDR Goes to War: How Expanded Executive Power, Spiraling National Debt, and Restricted Civil Liberties Shaped Wartime America*. New York: Threshold Editions, 2011.

Fontani, Marco, Mariagrazia Costa, and Mary Virginia Orna. *The Lost Elements: The Periodic Table's Shadow Side*. New York: Oxford University Press, 2014.

Ford, Kenneth. *Building the H Bomb: A Personal History*. Singapore: World Scientific, 2015.

French, Anthony, and P. J. Kennedy. *Niels Bohr: A Centenary Volume*. Cambridge, Mass.: MIT Press, 1985.

Frisch, Otto. *What Little I Remember*. Cambridge, England: Cambridge University Press, 1979.

Galison, Peter, and Bruce Hevly. *Big Science: The Growth of Large-Scale Research*. Stanford, Calif.: Stanford University Press, 1992.

Gamow, George. *Thirty Years That Shook Physics: The Story of Quantum Theory*. New York: Doubleday, 1966, reprinted in Dover Editions, 1985.

Glauber, Roy. "An Excursion with Enrico Fermi, 14 July 1954." *Physics Today*, June 2002, pp. 44–46.

Goodstein, Judith. *The Volterra Chronicles*. Providence, R.I.: American Mathematical Society, 2007.

Goudsmit, Samuel. *Alsos*. New York: Henry Schuman, 1947.

Groves, Leslie. *Now It Can Be Told: The Story of the Manhattan Project*. New York: Da Capo Press, 1962.

Guerra, Francesco, Matteo Leone, and Nadia Robotti. "Enrico Fermi's Discovery of Neutron-Induced Artificial Radioactivity: Neutrons and Neutron Sources." *Physics in Perspective* 8, 2006, pp. 255–81.

———. "Enrico Fermi's Discovery of Neutron-Induced Artificial Radioactivity: The Influence of His Theory of Beta Decay." *Physics in Perspective* 11, 2009, pp. 379–404.

Hahn, Otto. *My Life*. New York: Herder and Herder, 1970.

Heisenberg, Werner. "Quantum Theoretical Re-Interpretation of Kinematical and Mechanical Relations." *Zeitschrift für Physik* 33, 1925, pp. 879–93 (English translation in Van der Waerden, see below, p. 261).

———. *Physics and Beyond: Encounters and Conversations*. Translated by Arnold Pomerans. New York: Harper & Row, 1971.

———. *Quantum Theory and Measurement*. Princeton, N.J.: Princeton University Press, 1983.

Heilbron, John. *The Dilemmas of an Upright Man: Max Planck as Spokesman for German Science*. Berkeley: University of California Press, 1986.

Hermann, Armin. *Max Planck: The Genesis of Quantum Theory*. Cambridge, Mass.: MIT Press, 1971.

Holton, Gerald. "Striking Gold in Science: Fermi's Group and the Recapture of Italy's Place in Physics." *Minerva* 12, 1974, pp. 159–98.

Howes, Ruth, and Caroline Herzenberg. *Their Day in the Sun: Women of the Manhattan Project*. Philadelphia: Temple University Press, 1999.

Jackson, Allyn. *A Century of Mathematical Meetings*. American Mathematical Society, 1996, pp. 10–18.

Jacob, Maurice. *Giancarlo Wick: A Biographical Memoir*. Washington, D.C.: Proceedings of the National Academy of Sciences, 1999.

Kelly, Cynthia, ed. *The Manhattan Project*. New York: Black Dog & Leventhal, 2007.

Kertzer, David. *The Pope and Mussolini: The Secret History of Pope Pius XI and the Rise of Fascism in Europe*. New York: Random House, 2014.

Kevles, Daniel. *The Physicists: The History of a Scientific Community in Modern America*. Cambridge, Mass.: Harvard University Press, 1971.

Klein, Martin. *Paul Ehrenfest: The Making of a Theoretical Physicist*. New York: Elsevier, 1970.

Kragh, Helge. *Quantum Generations: A History of Physics in the Twentieth Century*. Princeton, N.J.: Princeton University Press, 1999.

Kronig, R., and Victor Weisskopf, eds. *Wolfgang Pauli: Collected Scientific Papers*, vols. 1 and 2. New York: Interscience, 1964.

Lanouette, William. *Genius in the Shadows: A Biography of Leo Szilard*. New York: Charles Scribner's Sons, 1992.

Laurence, William L. *Men and Atoms*. New York: Simon & Schuster, 1959.

Libby, Leona Marshall. *The Uranium People*. New York: Crane, Russack, 1979.

Maltese, Giulio. *Enrico Fermi in America*. Bologna: Zanichelli, 2003.

———. *Il Papa e l'Inquisitore*. Bologna: Zanichelli, 2010.

Marx, George. *The Myth of the Martians and the Golden Age of Hungarian Science.* Amsterdam: Kluwer, 1996.

McCullough, David. *Truman.* New York: Simon & Schuster, 1983.

Mehra, Jagdish. *The Physicist's Conception of Nature.* Dordrecht: Reidel, 1973.

Mehra, Jagdish, and Helmut Rechenberg. *The Historical Development of Quantum Mechanics.* Berlin: Springer Verlag, 1982.

Meitner, Lise. "Max Planck als Mensch." *Naturwissenschaften* 45, 1958, p. 406.

Moore, Walter. *Schrödinger: Life and Thought.* Cambridge, England: Cambridge University Press, 1989.

Nobel Lectures. Physics 1922–41, Amsterdam: Elsevier, 1965.

———. Physics 1942–62, Amsterdam: Elsevier, 1964.

———. Physics 1901–95. Singapore: World Scientific, 1996.

Noddack, Ida. "Über das Element 93." *Zeitschrift Angewandte Chimie* 47, 1934, pp. 653–55.

OHI (Oral History Interviews), American Institute of Physics, https://www .aip.org/history-programs/niels-bohr-library/oral-histories.

Oliphant, Mark. "The Beginning: Chadwick and the Neutron." *Bulletin of the Atomic Scientists,* December 1982, pp. 14–19.

Orear, Jay. *Enrico Fermi: The Master Scientist.* Internet First University Press, 2004.

Pais, Abraham. *Inward Bound.* New York: Oxford University Press, 1986.

———. *Niels Bohr's Times.* New York: Oxford University Press, 1991.

———. *The Genius of Science: A Portrait Gallery of 20th Century Physicists.* New York: Oxford University Press, 2000.

Pauli, Wolfgang. "Remarks on the History of the Exclusion Principle." *Science* 103, 1946, p. 213.

Pearson, J. Michael. "On the Belated Discovery of Fission." *Physics Today,* June 2015, p. 40.

Peierls, Rudolf. *Bird of Passage.* Princeton, N.J.: Princeton University Press, 1985.

Pontecorvo, Bruno. *Enrico Fermi.* Pordenone: Edizioni Tesi, 1993.

Rabi, Isidor. *Science: The Center of Culture.* New York: World, 1970.

Rasetti, Franco. Oral History Interview for Archives of the California Institute of Technology, 1982.

Rhodes, Richard. *The Making of the Atomic Bomb.* New York: Simon & Schuster, 1986.

———. *Dark Sun: The Making of the Hydrogen Bomb.* New York: Simon & Schuster, 1995.

Rigden, John. *Rabi: Scientist and Citizen.* New York: Basic Books, 1987.

Rosenfeld, Leon. *Nuclear Structure with Neutrons*. Amsterdam: North Holland, 1966.

Rotblat, Joseph. "Leaving the Bomb Project." *Bulletin of the Atomic Scientists*, August 1985, pp. 16–19.

Rotblat, Joseph, and Daisaku Ikeda. *A Quest for Global Peace: Rotblat and Ikeda on War, Ethics and the Nuclear Threat*. London: I. B. Tauris, 2006.

Rutherford, Ernest. "Collisions of Alpha Particles with Light Atoms." *Philosophical Magazine* 37, 1919, p. 581.

Schrödinger, Erwin. "Quantization as an Eigenvalue Problem." *Annalen der Physik* 79, 1926, pp. 361–76.

———. "On the Relation of the Heisenberg-Born-Jordan Quantum Mechanics to Mine." *Annalen der Physik* 79, 1926, pp. 734–56.

Schweber, Silvan. *QED and the Men Who Made It*. Princeton, N.J.: Princeton University Press, 1994.

———. *Nuclear Forces: The Making of the Physicist Hans Bethe*. Cambridge, Mass.: Harvard University Press, 2012.

Segrè, Claudio. *Atoms, Bombs and Eskimo Kisses*. New York: Viking, 1995.

Segrè, Emilio. *Enrico Fermi, Physicist*. Chicago: University of Chicago Press. 1970.

———. *From X-Rays to Quarks*. San Francisco: W. H. Freeman, 1980.

———. *A Mind Always in Motion*. Berkeley: University of California Press, 1993.

Serber, Robert. *The Los Alamos Primer*. Berkeley: University of California Press, 1992.

Sime, Ruth. *Lise Meitner: A Life in Physics*. Berkeley: University of California Press, 1996.

Smith, Alice Kimball. *A Peril and a Hope: The Scientists' Movement in America 1945–47*. Cambridge, Mass.: MIT Press, 1965.

Smyth, Henry. *Atomic Energy for Military Purposes*. Princeton, N.J.: Princeton University Press, 1945.

Sommerfeld, Arnold. *Atombau und Spektralinien*. Braunschweig: Vieweg Verlag, 1919. Translated by Henry L. Brose as *Atomic Structure and Spectral Lines*. New York: Dutton, 1923.

Steuwer, Roger. "Bringing the News of Fission to America." *Physics Today*, October 1985, pp. 49–56.

Strauss, Lewis. *Men and Decisions*. New York: Doubleday, 1962.

Szanton, Alexander. *The Recollections of Eugene Paul Wigner*. New York: Plenum Press, 1992.

Telegdi, Valentine. "Enrico Fermi in America." *Physics Today*, June 2002, pp. 38–42.

Trigg, George. *Landmark Experiments in Twentieth-Century Physics.* New York: Dover, 1995.

Truman, Harry. *Year of Decision.* New York: Doubleday, 1955.

Ulam, Stanislaw. *Adventures of a Mathematician.* New York: Charles Scribner's Sons, 1970.

United States Atomic Energy Commission (USAEC), *In the Matter of J. Robert Oppenheimer.* Cambridge, Mass.: MIT Press, 1954.

Van der Waerden, Bartel. *Sources of Quantum Mechanics.* Amsterdam: North Holland, 1967.

Von Baeyer, Hans. *The Fermi Solution: Essays on Science.* New York: Random House, 1993.

Von Hippel, Frank. "James Franck: Science and Conscience." *Physics Today*, June 2010, pp. 41–46.

Wattenberg, Albert. "The Building of the First Chain Reaction." *Bulletin of the Atomic Scientists*, June 1974, pp. 51–57.

———. "December 2, 1942: The Event and the People." *Bulletin of the Atomic Scientists*, December 1982, pp. 22–33.

Weart, Spencer, and Gertrude Weiss Szilard. *Leo Szilard: His Version of the Facts.* Cambridge, Mass.: MIT Press, 1978.

Weiner, Charles, ed. *History of Twentieth-Century Physics.* New York: Academic Press, 1977.

Weisskopf, Victor. *The Joy of Insight: Passions of a Physicist.* New York: Basic Books, 1991.

Wilson, David. *Rutherford, Simple Genius.* London: Hodder, 1983.

Wilson, Jane. "All in Our Time." *Bulletin of the Atomic Scientists*, March 1975, p. 35.

Wilson, Jane, and Charlotte Serber, eds. *Standing By and Making Do: Women of Wartime Los Alamos.* Los Alamos, N.M.: Los Alamos Historical Society, 1997.

Wilson, Robert. "The Conscience of a Physicist." *Bulletin of the Atomic Scientists*, June 1970, p. 30.

———. "A Recruit for Los Alamos." *Bulletin of the Atomic Scientists*, March 1975, p. 41.

Zinn, Walter. "Fermi and Atomic Energy." *Reviews of Modern Physics* 23, July 1955, pp. 263–68.

Zuccotti, Susan. *The Italians and the Holocaust.* New York: Basic Books, 1987.

ACKNOWLEDGMENTS

This book has been written with the support and encouragement of many people and we are in their debt. A word or two of heartfelt gratitude is much deserved.

First of all, thanks to Sir Christopher Llewellyn Smith, who in 1996 planted the original idea of a book about Enrico Fermi into Gino's mind, thinking undoubtedly that Italian physicists need to stick together. Then to the family historian, Olivia Fermi, who has kept her grandparents' legacy alive by an open cultural dialogue (see the Neutron Trail website). Her enthusiastic support of our book was also reflected in lengthy and often moving interviews with other members of the Fermi family: Sarah Fermi, the widow of Fermi's son, Judd (Giulio); and Rachel Fermi, Enrico and Laura's youngest granddaughter. Gino's cousin Fausta Walsby shared some childhood memories of the Fermi family from their overlapping time in Los Alamos with her parents, Emilio and Elfriede Segrè. And more background was generously provided by Robert Fuller, who regarded Judd Fermi as his best friend. Our meeting in Ithaca with Rose Bethe and her son Henry also provided invaluable anecdotes of the Fermi history.

We particularly want to acknowledge the kindness and insights of

Ugo Amaldi, himself a physicist, who spent the greater part of a day with us in Geneva, speaking about Fermi and his family, lifelong friends of his parents, Edoardo and Ginestra Amaldi. Several follow-up communications offered an even richer rendering of this friendship and a talk with Ugo's brother, Francesco, was helpful in filling a few blanks in the picture.

In Rome, we were delighted to have a personal tour of the Fermi museum at the university, as well as repeated help with our research from the historian Adele La Rana. Thanks also to the historian Giovanni Battimelli who offered us advice, guided us with his own writings, and provided precious access to other documents. We benefited in Pisa from the counsel of Roberto Vergara Caffarelli and the Domus Galilaeana's welcome, where the able assistance of Maura Beghè in gaining access to its Fermi Archives was greatly valued.

Special thanks also go to the courteous and more than competent archivists Diane Harper and Barbara Gilbert at the Regenstein Library of the University of Chicago and to Savannah Gignac at the Emilio Segrè Visual Archives of the American Institute of Physics. For the American part of this book, the trip that Ellen Bradbury Reid arranged for us a few years ago to Trinity, the site of the first atom bomb explosion, lent extra poignancy to this story.

Several esteemed physicists contributed astute observations and comments regarding Fermi and his collaborators. The list includes Harold Agnew, Jeremy Bernstein, Frank Close, Freeman Dyson, Kenneth Ford, Jerry Friedman, Richard Garwin, and Murray Gell-Mann. We are especially grateful to two distinguished physicists, Kenneth Ford and Alfred Goldhaber, who have read the manuscript in galley form, giving us valuable advice and pointing out errors. Needless to say, any remaining errors are purely our own fault.

We were lucky to have an editor, Serena Jones at Henry Holt, who was a strong believer in the importance of telling this story and guided the writing of it with a gentle and deft hand. And special thanks to Molly Bloom and Emily DeHuff, whose meticulous edits improved the manuscript. We are also grateful to Katinka Matson and John Brockman, literary agents who continue to find imaginative and creative ways of making the world of science accessible to a larger public.

And finally, many thanks to Doron Weber and to the A. P. Sloan Foundation's program in the Public Understanding of Science, Technology and Economics for providing the means to travel and absorb the many influences on Fermi's life. This book benefited enormously from their support.

INDEX

academic life and institutions, 22, 257
  fission research, 159–66, 174–204
  Institutes of Basic Research, 254–57
  Italian, 22–27, 28–30, 33–34, 40–48,
    74–75, 114, 120, 122
  Jews and, 34, 74–75, 86, 87, 118,
    127–28, 134
  Mussolini and, 74–75
  postwar, 254–57, 268–74
  United States, 76–77, 112, 118–20,
    136–37, 148–50, 154, 159–66,
    174–204, 254–57, 264, 268–74
  See also specific universities
Accademia dei Lincei, 48, 72, 73, 99, 287
Advisory Committee on Uranium,
    157–58, 162, 164, 173
Agnew, Harold, 181, 270, 284
Albuquerque, 208
Allison, Samuel, 171–72, 254, 255
alpha decay, 82–83
alpha particles, 82, 84, 91, 95, 103, 132,
    147
aluminum, 95, 96
Alvarez, Luis, 112, 145–46, 278
Amaldi, Edoardo, 67–69, 76, 84, 89–90,
    97–98, 102, 104, 108, 111, 112, 114,

119, 120, 122, 133, 154, 218, 252, 276,
    280, 285, 293, 294
Amaldi, Ginestra, 89, 104, 114–15, 120,
    122, 293, 294
Amaldi, Ugo, 89, 122, 294
American Institute of Physics, 273
American Physical Society, 288,
    291
Amidei, Adolfo, 14–17, 19, 20
Anderson, Carl, 112
Anderson, Herbert, 139–40, 142, 145,
    152, 163–66, 179, 180, 185, 202–4,
    251, 255, 265, 266, 297
  CP-1 experiment, 185–98
  Fermi and, 140, 185–98
Anschluss, 115, 128
antimatter, theory of, 91
Anti-Prossimo, 25–26
anti-Semitism, 34, 74–76, 86–88, 110,
    116–19, 121, 127, 134, 136, 146–48,
    154, 211, 216–18, 284–87, 305–6
Apuanian Alps, 25
Argentina, 100
Argonne, 182, 210, 212, 213, 265
art, 117, 271
Associated Press, 141

Association of Los Alamos Scientists
    (ALAS), 250–52
astrophysics, 272
atom, 22, 30, 40, 43–46, 69, 70
    Bohr model, 22, 43–44
    nucleus, 81–86
    wave function and, 50–52
atomic bomb, 133, 145
    aftermath of Japanese bombings,
        244–57, 261
    beginning of race for, 151–58
    chain reaction and, 144–50, 152–54,
        162–66, 178–83, 191
    CP-1 experiment, 184–204
    decision to use on Japan, 236–40
    detonation mechanism, 221–22
    ethics of, 240, 247–52, 256–57
    Fat Man, 243, 249
    fission and, 154–75, 176–243
    Franck report, 238–40
    Germany and, 133, 149, 151, 153,
        156–58, 172, 184, 197, 219–20, 225,
        235
    implosion method, 222, 225, 236,
        240
    Little Boy, 242–43
    Manhattan Project, 204–43
    MAUD Report, 170–72
    Met Lab project, 175, 176–204
    plutonium, 221, 222–25, 236, 240, 243
    postwar debates on, 250–52, 275–82
    public opinion on use of, 245–48
    Roosevelt policy on, 158–59, 173, 204,
        235
    secrecy and, 150, 163, 178, 184–85, 191,
        202, 204, 232, 234, 246–48, 249
    Smyth report, 248–49, 261
    Soviet Union and, 278, 282
    Trinity test, 1–3, 240–42, 247, 271, 307
    Truman policy, 235–43, 246, 248,
        251–52, 278
    U.S. research and funding, 151–58,
        162–75, 176–83, 184–243
    used on Japan, 242–43, 244–52
    U-235, 221, 222, 236, 240, 242–43
atomic energy, 197, 203
    birth of, 193–98
    positive uses of, 197, 250–51, 256
    postwar debates on, 250–52, 256–57,
        261–67, 275–82, 288–93

Atomic Energy Act, 275
Atomic Energy Commission (AEC),
    276–90, 302
attraction, 82
Auschwitz, 218
Austria, 17, 35, 53, 100
    annexation of, 115, 128
    World War I, 17
automobiles, 59–60, 176

Badoglio, Pietro, 107, 113
Bainbridge, Kenneth, 1
Bari, 10
barium, 96, 129, 137
Baudino, John, 212, 226, 246
Belgium, 52, 53, 119, 156
    World War II, 161
Bell Laboratories, 231
Berkeley, 112, 113, 145, 166, 168, 169, 174,
    206, 207, 211, 264, 289, 306
Berlin, 84, 94, 105, 111, 115, 127, 128, 129,
    146
Berlin-Dahlem Institute, 84, 94, 105, 158
beryllium, 84, 95, 128, 140, 152, 153
beta decay, 82–83, 88–94, 109, 168, 263
    Fermi theory on, 88–94, 109, 263
Bethe, Hans, 2, 78, 88, 90, 91, 176, 207,
    225, 226, 257, 278, 281, 289
Bethe, Rose, 176
Bhagavad Gita, 2
Bloch, Felix, 80, 88, 207
Bohr, Harald, 139
Bohr, Niels, 22, 39, 43–44, 52, 54, 80, 82,
    111, 119, 129–34, 136–43, 231, 235,
    250, 264, 285
    complementarity principle, 52
    Fermi and, 141–43
    fission and, 129–34, 137–44, 150
    in Los Alamos, 213–32
    model of atom, 22, 43–44
    in United States, 136–43, 231–32
Bohr Institute, Copenhagen, 127, 137, 138
Bologna, 7, 29
Born, Max, 30, 32, 49, 50, 87
boron, 153, 162
Bothe, Walter, 84, 163
Boys of Via Panisperna, 67–69, 76–77, 83,
    85–86, 94–109, 110–11, 117, 122, 152,
    154, 215, 229, 285, 297
    breakup of, 107–9

Proceedings paper, 105–7
rise and fall of, 102–9
transuranics and, 98–99, 105–7,
    122–23, 130, 132–33
Bradbury, Norris, 276
Brazil, 100
Briggs, Lyman, 157, 164, 165, 170–71,
    173
Bristol University, 265
British Mission, 220–21, 234
British Secret Service, 280
Brussels, 52, 88
Bryson, William C., 244
Buck, Pearl, 123
Budapest, 147
Buenos Aires, 100
Bush, Vannevar, 164–65, 167, 170–74, 181,
    236
Byrnes, James, 236, 237, 238, 284

cadmium, 190
cadmium sulfate, 194
Caltech, 76, 77
Cambridge, 31, 111, 301
Campo dei Fiori, Rome, 15–16
Canada, 154, 220, 248
cancer, 256, 296–97
Caorso, 8–10
Capon, Augusto, 56–57, 67, 88, 114–15,
    161, 216, 217, 218, 284, 286
    death in Auschwitz, 218
Capon, Laura. *See* Fermi, Laura
Caraffa, Andrea, *Elementorum Physicae
    Mathematica*, 16
carbon, 153, 162, 164, 220
Carelli, Antonio, 117
Carnegie Institution, 138, 140, 157, 164
Carrara, 25
Carrara, Nello, 26
Castelnuovo, Gina, 57–58, 60, 287
Castelnuovo, Guido, 33, 34, 57, 67, 287
Catholicism, 10, 62, 74, 122
Cavendish Laboratory, 111
celestial motion, mechanics of, 16
CERN, 293
Chadwick, James, 85, 90, 94, 221, 263
    discovery of neutron, 85
chain reaction, 144–50, 152–54, 162–66,
    174, 178–83, 191, 249
    patent, 147, 148

Chandrasekhar, Subrahmanyan, 272, 297,
    298
chemistry, 33, 133, 134
Chicago, 174–75, 176–204
    Back of the Yards, 187, 190
    Fermi's postwar life and work in,
        253–57, 261–302
    Met Lab project, 175, 176–204
China, 270
Churchill, Winston, 135, 165, 241
Ciano, Galeazzo, 159
cloud chamber, 84
Cold War, 275, 277, 280–82, 288
Collegio Romano, 16
Columbia River, 205–6
Columbia University, 97, 119, 154, 174,
    187, 254, 255, 264, 306
    Fermi at, 119–20, 136–37, 140, 142, 145,
        152, 159–66, 179
Communism, 75, 207, 284, 286
    McCarthyism and, 288–90, 291, 302
Como, 53, 61, 66, 285–86, 292
complementarity principle, 52
Compton, Arthur, 77, 164, 167–75, 178,
    180, 184, 188–89, 194–98, 207, 212,
    222, 223, 236–40, 255, 302
Compton, Karl, 164
computers, 55, 266, 293
Conant, James, 164, 171, 173–74, 181, 197,
    222, 236, 278, 279
concentration camps, 132, 211, 217–18
Concordato, 74
Consiglio Nazionale delle Ricerche,
    47–48, 97, 113
Copenhagen, 52, 80, 111, 119, 127, 131,
    137, 138, 141, 150, 231
Copenhagen Interpretation of Quantum
    Mechanics, 52, 54
Copernicus, Nicolaus, 122
Corbino, Orso Mario, 29–30, 33, 38, 41,
    42, 45, 47, 48, 53, 59, 61, 62, 65,
    67–69, 73, 77, 81, 95, 98–99, 104, 106,
    110–11, 285
    death of, 110
Cornell University, 150, 273
Corriere della Sera, Il, 120–21
cosmic rays, 154, 265, 285, 292
CP-1 experiment, 184–204, 205, 222
CP-2 experiment, 203–4, 205, 213,
    222

Curie, Irène, 84–85, 90–91, 94, 95, 96, 98,
    105, 106, 108, 133
Curie, Marie, 94, 133, 186, 272
Curie-Joliot experiment, 84–85, 90–91,
    94, 95, 96, 98
cyclotron, 112–14, 139, 140, 145, 147, 154,
    168, 263, 264–65, 293
Czechoslovakia, 80, 156
    German invasion of, 151

Dachau, 132
Daghlian, Haroutune, 233
D'Agostino, Oscar, 97, 104, 108
Dante, *Divina Commedia,* 14
Denmark, 54
deuterium, 163
Development of Substitute Materials
    Program, 204
Diebner, Kurt, 158
Difesa della Razza, 118
Dirac, Paul, 31, 32, 50–52, 69, 78,
    264
    Fermi-Dirac statistics, 51–52
    "The Fundamental Equations of
    Quantum Mechanics," 50
Dolomites, 35, 56, 57, 59, 67, 68, 89, 100,
    293
Dresden, 238, 245
Dunkirk, 161
Dunning, John, 138, 140, 142
DuPont chemical company, 183, 190–91,
    195–96, 223, 224, 256
Dyson, Freeman, 273

Ehrenfest, Paul, 39, 78
Einstein, Albert, 22, 27, 39–40, 51, 53, 87,
    105, 130, 146–47, 156–57, 255, 261,
    264
    atomic bomb and, 156–57
    Fermi and, 39–40
    theory of relativity, 22, 27, 264
    in United States, 87
electric battery, 53
electric forces, 27, 53, 82–84, 90, 93, 112
electron microscope, 147
electrons, 43–46, 82–83, 89, 98, 111
    orbits, 43–46, 49, 50
    wave function and, 50–52
Enciclopedia Treccani, 70
Enola Gay, 145, 242

Enrico Fermi Award, 302
Enrico Fermi International School of
    Physics, 292
Enrico Fermi Prize, 268
Enriques, Federico, 33, 34, 74, 122
Ethiopia, Italian invasion of, 107–8
eugenics, 86
Exclusion Principle, 46–47, 55
experimental physics, 32, 43–46, 84,
    93–99, 102–7, 112–14, 140–43, 196,
    263, 264, 269
extraterrestrial life, 273–74

Fano, Ugo, 80, 119
Fascism, 36–38, 47–48, 72–75, 88, 99, 106,
    110, 115, 216–17, 284
    fall of, 216–17
    Fermi and, 74–75, 88, 116, 123–24,
    161
Fascist Grand Council, 216
Fat Man, 243, 249
Federation of American Scientists, 252
Federzoni, Luigi, 123–24
Fermi, Alberto, 8–12, 13–17, 56
    death of, 56
Fermi, Enrico, 2–3, 10
    academia and, 22–30, 33–34, 40–48,
    114, 118–22, 136, 140, 154, 254–57,
    268–74
    on Advisory Committee on Uranium,
    157–58, 162, 164
    aftermath of Japanese bombings,
    244–57, 261
    Americanization of, 159–62, 175, 287
    ancestral roots, 7–12
    atomic bomb development and,
    171–75, 176–83, 184–243
    atomic bomb used on Japan and,
    242–43, 244–52
    becomes U.S. citizen, 213–14
    beta decay theory, 88–94, 109, 263
    birth of, 11
    Bohr and, 141–43
    chain reaction and, 144–50, 152–54,
    162–66, 178–83, 191
    childhood of, 11–17
    choice of physics, 18–27
    Collected Works, 40
    at Columbia University, 119–20, 136–37,
    140, 142, 145, 152, 159–66, 179

CP-1 experiment, 184–204
death of, 302
discovery of neptunium, 98–99, 106, 122
"Distinctive Characteristics of Sound and Their Causes," 19–20
education of, 14–28
Einstein and, 39–40
emotional control of, 12, 15, 35, 241, 295
as enemy alien, 177–78, 192, 213
as experimentalist, 32, 43–46, 59, 84, 93–99, 102–7, 112–14, 140–43, 196, 263, 264
fame of, 79, 81, 94–95, 254–55, 261–63, 301
as Henry Farmer, 212–16
Fascist regime and, 74–75, 88, 116, 123–24, 161
as a father, 79, 109, 202, 295, 301
finances of, 71–73, 88, 104, 120, 283–84
La Fisica Moderna, 70–71
fission and, 137–43, 144–58, 161–83
"The Fission of Uranium," 142
Florence professorship, 41–47
GAC and, 276–80, 283, 286
Göttingen position, 30–32
Hanford project and, 204, 210, 212–14, 221, 222–25
hydrogen bomb and, 278–82
immigration to U.S., 118–24, 135–36, 155
Institutes of Basic Research and, 254–57
Introduzione alla Fisica Atomica, 71
in Leiden, 39–40
"Little Match" nickname, 13
in Los Alamos, 204, 211–18, 222, 224–43, 284, 287–89, 297, 306
Manhattan Project and, 204–43
marriage to Laura, 61–62, 71–72, 78–79, 100, 109, 122, 176–77, 185, 202, 226, 246–49, 285, 293–95, 300
Met Lab project, 175, 176–204
Molecole e Cristalli, 77
move to Chicago, 176–83
Mussolini and, 73–74, 116, 123–24, 161
Nobel Prize of, 95, 119–23, 132, 151, 262, 294

personality of, 12, 15, 35, 180, 230, 241, 271, 279, 289, 295
physical appearance of, 13, 25, 58, 61, 151, 296
in Pisa, 18–27
political views, 37–38, 48, 73, 75, 87–88, 161, 172, 276, 286
"Pope of Physics" nickname, 3, 68, 79
postwar life and work, 253–57, 261–302
quantum field theory and, 78, 263
radioactivity experiments, 94–99, 102–6
Franco Rasetti and, 24–26, 42–45, 60, 68, 77, 96–97, 248, 293–94
return to Chicago, 253–57
return to Italy, 278, 283–87
Ricerca papers, 96–97
rise to prominence, 69, 70, 79, 81, 94
Rome professorship, 47–48, 56, 65–69
in Royal Italian Academy, 72–75
scattering experiments, 265–66, 273
on Scientific Panel, 236–40
at Scuola Normale Superiore, 16–17, 18–28
secrecy of war effort and, 178, 184–85, 204, 232, 234, 249
on slow neutrons, 102–6, 152–54
statistical mechanics and, 51–52, 54, 69
stomach cancer of, 296–97
Szilard and, 149–54, 162, 255–56
teaching style of, 67–68, 269–74
"The Theory of the Collisions between Atoms and Electrically Charged Particles," 40
thesis of, 26–28
Thomas-Fermi equation and, 69
transuranics and, 98–99, 105–7, 122–23, 130, 132–33
Trinity test and, 1–3, 240–42, 271, 307
World War II and, 176–83, 211–43, 262
writings of, 26–27, 32, 40, 46–47, 70–72, 77–78, 82, 88–90, 96–97, 104–8, 142, 268–69
Fermi, Giulia, 8–10
Fermi, Giulio (brother of Enrico), 11, 13, 15, 16, 18, 109
Fermi, Giulio (Judd, son of Enrico), 109, 122, 135, 154, 160, 176–77, 202, 214, 215, 226–28, 253–54, 284, 294–95, 301

Fermi, Ida, 10–12, 13–17, 35
    death of, 35–36
Fermi, Laura, 56–62, 68, 89, 98, 104, 108,
    114–15, 116, 129, 154, 212, 241, 242,
    307
    Alchimia del Nostro Tempo, 114
    Americanization of, 160–61, 176–77
    Atoms in the Family, 294, 296,
        298–99
    becomes a U.S. citizen, 213–14
    death of, 300
    death of Fermi, 299–302
    Holocaust and, 217–18
    immigration to U.S., 118–24, 135–36,
        155
    Judaism of, 56–59, 116–19, 123, 136,
        216–18, 306
    in Los Alamos, 213–18, 226–33
    marriage to Fermi, 61–62, 71–72,
        78–79, 100, 109, 122, 176–77, 185,
        202, 226, 246–49, 285, 293–95,
        300
    move to Chicago, 176–77
    return to Italy, 283–87
    writing career, 294, 296, 298–99
Fermi, Maria, 11, 13, 14, 56, 62, 68, 217,
    218, 247–48, 294
Fermi, Nella, 79, 100, 122, 135, 154, 160,
    176–77, 202, 214, 215, 226–29, 254,
    284, 295, 301
Fermi, Olivia, 301
Fermi, Stefano, 8–10
Fermi coordinates, 27
Fermi-Dirac statistics, 51–52, 55
Fermilab, 256–57
Fermi method, 270–74
Fermi questions, 270
Fermi sea, 51
Fermi surface, 51
Ferrovie dello Stato, 10
Feynman, Richard, 78, 230–31
fishing, 229–30
fission, 127–58, 162–83, 249, 277
    atomic bomb and, 144–75, 176–243
    as chain reaction, 144–50, 152–54,
        162–66, 178–83, 191
    discovery of, 129–34, 136–43
    Hahn-Strassmann experiment,
        129–34, 137, 142
    Met Lab project, 175, 176–204
    terminology, 141
    U.S. research and funding for atomic
        bomb, 151–58, 162–75, 176–83,
        184–204
Florence, 41–47, 59, 117
Ford, Ken, 281
France, 53, 57, 107, 153, 291
    World War I, 17
    World War II, 159, 161
Franck, James, 87, 238–40
Franck report, 238–40
Franco, Francisco, 115
Franconia, RMS, 135, 148
Frisch, Otto, 127–34, 145, 150, 170, 221,
    230
    discovery of fission, 129–34, 136–43
Fuchs, Klaus, 234, 280
fusion, 277, 282
Fussell, Paul, 246

Gadget, 211, 214, 234, 241, 249–50
Galileo, 20, 42, 122, 307
Gamow, George, 138
Garbasso, Antonio, 41, 42, 43, 47
Garwin, Richard, 298
gas masks, 187
Geiger counters, 162
Gell-Mann, Murray, 298
General Advisory Committee (GAC),
    276–80, 283, 286
general relativity, 22, 27, 41
Geneva Protocol, 107
Genoa, 62
geometry, 14, 19, 27, 269
George Washington University, 138, 140,
    149
Germany, 23, 30–32, 37, 48, 53, 84, 100,
    107
    annexation of Austria, 115, 128
    anti-Semitism and racial laws, 86, 87,
        117, 118, 121, 127–28, 134, 147–48,
        154, 211
    atomic bomb and, 133, 149, 151, 153,
        156–58, 172, 184, 197, 219–20, 225,
        235
    Fermi and, 30–32
Gestapo, 37
Jews, 86, 87, 117, 121, 127–28, 154, 155,
    170, 238
Kristallnacht, 121, 132

Nazi, 86–88, 100, 115, 117, 121, 127, 128, 147–49, 154, 158, 170, 184, 211, 216–18, 286
  physics in, 30–32, 49–52, 86, 87, 105–7, 111, 127–34, 158, 170, 219–20, 245, 264
  SS, 216, 217
  Uranverein, 158, 219–20
  World War I, 17
  World War II, 155, 159, 161, 173, 177, 211, 216–20, 225, 235, 238
  youth movement, 155
Goldberger, Marvin, 88–89
Goodyear Company, 190
Göttingen, 30–32, 39, 49–50, 51, 87, 111
Goudsmit, Sam, 78
graphite, 153, 158, 162–66, 179
  CP-1 experiment, 184–204
  -uranium experiment, 163–66
gravity, 18, 27, 43, 82, 89, 93
Great Britain, 53, 107, 111, 120, 148, 153, 165, 220–21, 301
  fission research, 169–72, 220–21
  World War I, 17
  World War II, 135, 159, 161, 165, 241
Greenewalt, Crawford, 195–96, 197, 223, 256
Groves, Leslie, 1–2, 182–83, 189, 195, 262
  Manhattan Project and, 204–8, 209–12, 222, 229, 236, 241–42, 248, 249
Gustavus V, King of Sweden, 122

Hahn, Otto, 84, 105, 106, 123, 127–34, 245
  discovery of fission, 129–34, 137, 142
  transuranics and, 106
Hahn-Strassmann experiment, 129–34, 137, 142
Hanford, 190–91, 204–6, 210, 212–14, 221, 222–25, 236, 249, 255, 256
Harper's Magazine, 246
Harteck, Paul, 220
Harvard University, 231
heat, 32–33
heavy water, 163, 220
Heisenberg, Werner, 31, 32, 49–52, 54, 78, 80, 85, 86, 154–56, 158, 220, 244, 245, 264
  matrix mechanics, 49–52

Nazi regime and, 155–56, 158, 245
  uncertainty principle, 52, 83
  unified field theory, 264
  in United States, 154–55
helium, 44, 82, 83, 153
Higgs boson, 293
hiking, 43, 58, 66, 67, 68, 229, 230, 292, 295, 306
Hilbery, Norman, 194
Hiroshima, 145, 242
  aftermath of bombing, 244–57, 261
  atomic bomb dropped on, 242–43, 244–52
Hitler, Adolf, 34, 74, 87, 100, 115, 123, 128, 133, 155, 161, 175, 177, 211, 219, 225
  anti-Semitism and racial policies of, 86, 87, 117, 118, 128
  death of, 235
  Mussolini and, 115, 117–18
  rise to power, 128, 147, 148
Hoerlin, Herman, 306
Hoerlin, Kate, 306
Holland, 128
  World War II, 161
Holocaust, 211, 217–18
Hooper, Stanford, 151
Hungary, 80, 146, 147, 274
  anti-Semitism in, 146
Hutchins, Robert, 189, 254–57
Huxley, Aldous, Brave New World, 108
hydrogen, 44, 82, 83, 98, 103, 104, 153, 163
hydrogen bomb, 266, 274, 277–82, 299
  test, 282

independent findings, 69
Industrial Revolution, 9
inflation, 37, 57
influenza, 19
Institute for Nuclear Studies, 133, 254, 255
Institute of Metallurgy, 254
Institute of Radiobiology and Physics, 254, 256
Institutes of Basic Research, 254–57
intergalactic magnetic fields, 265, 272
Interim Committee, 236–40
International Conference on Peaceful Uses of Atomic Energy (1955), 300

International Education Board, 38–39
Iowa State University, 186
Italian Physical Society, 27
Italian Society for the Advancement of
    Science, 113
Italy, 7–12, 53
    academic life and institutions, 22–27,
        28–30, 33–34, 40–48, 74–75, 114,
        120, 122
    anti-Semitism in, 34, 74–75, 87–88,
        116–19, 136, 216–18, 284–87,
        305–6
    currency, 57
    education, 14–28
    Ethiopian campaign, 107–8
    Fascism, 36–38, 47–48, 72–75, 88, 99,
        106, 110, 115, 216–17, 284
    Fermi returns to, 278, 283–87
    inflation, 37, 57
    Jews, 34, 56–59, 74–75, 87–88, 116–19,
        121, 123, 136, 161, 216–18, 226,
        284–87, 305–6
    physics in, 18–27, 31–34, 41–48, 52, 54,
        66, 67, 70, 79–86, 93–114, 119, 161,
        261, 285–87, 293
    postwar, 284–87, 306
    press, 83, 99, 118, 120–21, 123
    propaganda, 38, 54, 99, 106, 108
    railroads, 9, 10, 14
    unification, 9, 11, 65, 74, 88
    World War I, 17, 19, 29, 34, 37, 107
    World War II, 159, 161, 173, 216–18

Japan, 92, 233
    atomic bombs dropped on, 242–43,
        244–52
    attack on Pearl Harbor, 173–74, 202,
        237
    physics in, 264–65
    World War II, 173–74, 235–43
Jewett, Frank, 164
Jews, 34, 74, 78–79
    academia and, 34, 74–75, 86, 87, 118,
        127–28, 134
    American, 134
    anti-Semitism and, 34, 74–75, 86–88,
        110, 116–19, 127, 134, 136, 146–47,
        154, 211, 216–18, 284–87, 305–6
    German, 86, 87, 117, 154, 155, 170,
        238

Holocaust, 211, 217–18
    Italian, 34, 56–59, 74–75, 87–88,
        116–19, 121, 123, 136, 161, 216–18,
        226, 284–87, 305–6
Joe-1, 278
Joint Committee on Atomic Energy,
    280–81, 282
Joliot, Frédéric, 84–85, 90–91, 94, 95, 96,
    98, 105, 108, 153
Jordan, Pascual, 49, 78

k, 163, 166, 185, 221
Kaiser Wilhelm Gesellschaft, 48
Kaiser Wilhelm Institute for Chemistry,
    128
Khvolson, Orest, 18–19
Kistiakowsky, George, 1, 2
Kodoma, Michiko, 244
Korean War, 282
Kristallnacht, 121, 132
Krupp steel foundries, 115
Kyoto, 242

Large Hadron Collider, 293
lasers, 55
Lateran Pacts, 74
Latin, 16
Laval University, 154
Lawrence, Ernest, 77, 112, 145, 168–71,
    207, 236, 239, 278
League of Nations, 106, 115
Lee, Tsung Dao, 270
Leiden, 39–40, 76
Leipzig, 31, 80, 84, 86
Leonia, New Jersey, 160–61, 176, 283,
    284
Levi-Civita, Tullio, 27, 33, 34, 45, 122
Life magazine, 234, 299
Lilienthal, David, 277–78
lithium, 153
Little Boy, 242–43
littori, 106
London, 147
Los Alamos, 204, 206–8, 209–16, 226–43,
    249, 253, 297, 306
    aftermath of Japanese bombings, 244–52
    Bathtub Row, 215
    "Big House," 209
    Fermi in, 204, 211–18, 222, 224–43,
        284, 287–89, 297, 306

the Hill, 226–33
hydrogen bomb research, 278–82
life in, 226–33
nuclear stockpile, 275–76
postwar, 255, 275–83, 287–89, 306, 307
spies, 234, 280
wives, 226–33, 246–47
See also Manhattan Project; *specific scientists*
Los Alamos Primer, 211
Los Alamos Scientific Laboratory (LASL), 211, 306
Lo Surdo, Antonio, 47, 73, 111

magnetism, 47
Majorana, Ettore, 69, 76, 85–86, 87, 114, 117
disappearance of, 117
paranoia and isolation of, 86, 114
Manchuria, Soviet invasion of, 240
Manhattan Project, 204–43, 276
aftermath of Japanese bombings, 244–57, 261
bomb used on Japan, 242–43, 244–52
British Mission, 220–21, 234
choice of site, 204–8
decision to use bomb on Japan, 236–40
secrecy, 204, 232, 234, 246–48, 249
Smyth report, 248–49, 261
Trinity test, 1–3, 240–42, 247, 271, 307
Manifesto degli Scienziati Razzisti, 118
March on Rome, 37–38, 48, 72
Marconi, Guglielmo, 48, 54, 113
Marshall, George, 236
Marshall, John, 180, 213, 255, 265
Marshall, Leona. *See* Woods Marshall, Leona
mass, 82, 85, 130
mathematics, 14, 16, 19, 22, 33, 34
matrix mechanics, 49–52
Matteotti, Giacomo, 36, 48
Matterhorn, 66, 285
MAUD Committee, 170–72
report, 170–72
Mayer, Maria Goeppert, 272
May-Johnson bill, 251–52
McCarthyism, 288–90, 291, 302
McKibbin, Dorothy, 214
McMahon, Brien, 252, 280

McMillan, Edwin, 112, 168
medical technology, 256–57
Meitner, Lise, 84, 94, 105, 106, 123, 127–34, 145, 158
discovery of fission, 129–34, 136–43
transuranics and, 105, 106, 123
mesons, 264–65
Messina, 29, 47
Met Lab, 175, 176–204, 238, 248, 254
Committee on Political and Social Problems, 238
CP-1 experiment, 184–204
secrecy, 184–85, 191, 202, 204
Milan, 7, 10, 47, 286
Missouri, USS, 243
MIT, 164, 168, 210, 264
Radiation Laboratory, 210
Monsanto, 256
Montel, Alberto, 217
Montel, Anna, 217
Morocco, 97
Moscow, 177
motion, laws of, 20–21
MRIs, 55
Munich, 30, 80, 111, 115
Murrow, Edward R., 201
music, 10–11, 101, 219, 271
Mussolini, Benito, 36–38, 48, 57, 61, 72–74, 83, 88, 100, 106, 107, 108, 115–18, 154, 159, 175, 177, 284, 287
academia and, 74–75
anti-Semitic campaign, 118–19
ascent to power, 37–38, 72
Ethiopian campaign, 107–8
fall of, 216
Fermi and, 73–74, 116, 123–24, 161
Hitler and, 115, 117–18
science and, 53–54, 114
use of propaganda, 38, 54, 99, 106, 108
World War II and, 161
Mussolini Prizes, 72–74
mustard gas, 107

Nagasaki, 261
atomic bomb dropped on, 243, 244–52, 277
aftermath of bombing, 244–57, 261
Napoleon Bonaparte, 8, 177

National Academy of Sciences (NAS), 167, 169, 171, 173
National Accelerator Laboratory, 256–57
National Carbon Company, 162
National Defense Research Council (NDRC), 164–65, 167–75, 222
National Recovery Administration, 156
National Research Council, 48, 70
National Security Council, 278
Nature, 45, 90, 91, 99, 131–32, 137, 141, 142, 153
Naturwissenschaften, 137
Nazism, 86–88, 100, 115, 117, 121, 127, 128, 147–49, 154, 158, 170, 184, 211, 216–18, 286
neptunium, 168
    discovery of, 98–99, 106, 122
    naming of, 168
Netherlands, 39–40, 53
neutrinos, 90–92, 114
neutrons, 85–94, 102–6, 109, 111, 263
    beta decay theory, 88–94, 109, 263
    discovery of, 85–86
    slow, 102–6, 123, 127, 129–34, 150, 152–54, 163, 170, 172, 263, 285–86
New Jersey, 160, 283, 284
New Mexico, 204, 208–16, 226–33, 253–54
    Trinity test, 1–3, 240–42, 247
New York, 119–20, 135, 148
New Yorker, 296
New York Herald Tribune, 123
New York Times, 99, 121, 141, 249, 251, 298
Nobel Prize, 31, 46, 48, 53, 85, 94, 119–23, 133–34, 167, 207, 269
    of Fermi, 95, 119–23, 132, 151, 262, 294
    prejudice against women, 133–34, 272
Noddack, Ida, 106
nuclear medicine, 256–57
nuclear physics, 81–86, 111–15
    beta decay theory, 88–94, 109, 263
    chain reaction, 144–50, 152–54, 162–66, 178–83, 191
    CP-1 experiment, 184–204
    discovery of fission, 129–34, 136–43
    discovery of neutron, 85–86
    fission, 129–58, 162–83
    Manhattan Project, 204–43

Met Lab project, 175, 176–204
    postwar, 250–52, 256–57, 261–67, 275–82, 288–93
    U.S. research and funding for atomic bomb, 151–58, 162–75, 176–83, 184–243
    See also specific areas, scientists, and theories
nuclear reactors, 179, 192
    CP-1, 184–204
    CP-2, 203–4, 205
    Hanford, 212–14, 221–25, 236
nucleus, 43, 81–86, 112, 263
    discovery of, 82
    discovery of fission, 129–34, 136–43
    fission, 129–58, 162–83
Nuovo Cimento, Il, 27

Oak Ridge, 204–6, 210, 212, 221, 236, 249, 255, 256
Office of Demography and Race, 118
Office of Scientific Research and Development (OSRD), 173, 181, 182
Office of the Inquiry into the Social Aspects of Atomic Energy, 256
Oliphant, Marcus, 169–71
Oppenheimer, J. Robert, 1, 2, 77, 146, 147, 206–8, 236, 261, 276, 299
    communist affiliations, 288–91
    Manhattan Project and, 206–8, 210–11, 220–25, 229, 232, 239–41, 247, 249
    McCarthy hearings, 288–91
    opposition to H-bomb, 281–82, 288
    postwar nuclear debate and, 275, 276, 281–82, 288
Oppenheimer, Kitty, 229
Organization for Vigilance and Repression of Anti-Fascism (OVRA), 36–37
Oro alla Patria, 108
Ossietsky, Carl von, 119

Padua, 29
Pact of Steel, 159
Palermo, 107–8
paraffin, 102–4
Paris, 84, 108, 291
    World War II, 161
particle accelerators, 256–57, 263, 264–66, 292–93. *See also* cyclotron

patents, 104, 147, 148
  slow neutron, 285–86
Pauli, Wolfgang, 31, 32, 46, 50, 54, 78, 80,
    88, 89–90, 264
  Exclusion Principle, 46–47, 55
Pauli Effect, 32
Pauli Principle, 46–47, 55
Pauling, Linus, 77
Pearl Harbor, attack on, 173–74, 202, 237
Pegram, George, 136, 151, 169
Peierls, Genia, 215
Peierls, Rudolf, 79, 88, 90, 91, 170, 215,
    220, 221, 225, 226
Pentagon, 182, 236
Periodico di Matematiche, 70, 81
periodic table, 44, 82, 84, 95–99, 105, 153
  discovery of neptunium, 98–99, 106, 122
  transuranics, 98, 105–7, 122–23, 130,
    132–33, 168–69
  See also specific elements
Persico, Enrico, 15, 18, 19, 21, 23, 30, 31,
    39–41, 45, 47, 49, 75, 114, 248, 295
Philippines, 242
photons, 84–85, 264
Physical Review, 108, 142
physics, 15–16, 18
  anti-Semitism and, 86, 87, 127, 134,
    146–48, 154
  beta decay theory, 88–94, 109, 263
  discovery of fission, 129–34, 136–43
  fission, 129–58, 162–83
  German, 30–32, 49–52, 86, 87, 105–7,
    111, 127–34, 158, 170, 219–20, 245, 264
  Italian, 18–27, 31–34, 41–48, 52, 54, 66,
    67, 70, 79–86, 93–114, 119, 161, 261,
    285–87, 293
  Japanese, 264–65
  postwar, 250–52, 256–57, 261–67,
    275–82, 288–93
  in United States, 76–77, 105, 112–13,
    135–243, 254–57, 264–74
  Volta Congress, 52–55
  *See also specific areas, scientists, and
    theories*
piles, 163–66, 172, 173, 178–79, 182, 183
  CP-1 experiment, 184–204
pions, 265–66, 273
Pisa, 16–27, 293, 307
  Fermi in, 18–27
  Leaning Tower, 21

Pittarelli, Giulio, 20
Pius IX, Pope, 9, 74
Placzek, George, 80, 88, 150, 153
Planck, Max, 22, 43–44, 51
  concept of quanta, 44
plutonium, 94, 106, 122, 169, 172–74, 179,
    181, 221, 222
  bomb, 221–25, 236, 240, 243
  naming of, 169
  production, 190–91, 195–96, 205–6,
    221, 222–25, 236
pogroms, 121
Poincaré, Henri, *Théorie des Tourbillons,*
    22–23
Poland, 159, 224
  German invasion of, 159
polonium, 95, 128
Pontecorvo, Bruno, 102, 104, 108, 285, 286
Pontremoli, Aldo, 114
positrons, 91
Potsdam Conference, 241
Po Valley, 7, 8, 9
press, 83, 99, 131
  anti-Semitic, 118
  on discovery of fission, 141–43
  Italian, 83, 99, 118, 120–21, 123
  Scandinavian, 141
  United States, 99, 121, 123, 141, 160
  See also specific publications
Princeton University, 136, 137, 138,
    148–49, 174, 248, 264
  Institute for Advanced Study, 136, 261,
    281, 288
*Proceedings of the Cambridge
    Philosophical Society,* 69
*Proceedings of the Royal Society of
    London,* 51, 105
protons, 82–85, 89, 111, 129, 256, 263
Puccianti, Luigi, 22
Pu-239, 222, 223
Pu-240, 222

quanta, concept of, 44
quantum field theory, 78, 263, 264
quantum physics, 22, 30, 41, 43–46, 48,
    49–55, 77–78, 112, 167, 294
  Exclusion Principle, 46–47, 55
  rise of, 49–55
  *See also specific areas, scientists, and
    theories*

Quebec, 154
Quick and the Dead, The (documentary),
    262

Rabi, Isidor, 2, 77, 97, 134, 149, 210, 211,
    255, 279
Racah, Giulio, 80, 119
radiation, 40, 43, 44, 77, 84, 91
radiation poisoning, 94, 186, 233, 238,
    245, 297
radioactivity, 82–85, 94–99, 102–6,
    128
  induced, 102–3, 162
radio frequencies, 45
Radiology, 256
radiosodium, 112
radium, 94, 95–96, 128, 152
Radium Institute, 95
radon, 140, 152
Raphael, 72
Rasetti, Franco, 24–26, 41–43, 47, 50, 51,
    54, 61, 66–68, 76, 84, 94, 104, 107,
    112, 120, 122, 128, 154, 285
  at Caltech, 76–77
  in Canada, 154
  Fermi and, 24–26, 42–45, 60, 68, 77,
    96–97, 248, 293–94
relativity, 22, 27, 41, 44, 264
Renaissance, 8
Rendiconti dell'Accademia dei Lincei, 27
Resphighi, Ottorino, 101
Reviews of Modern Physics, 78
Ribbentrop, Joachim, 159
Ricerca Scientifica, La, 96–97, 104, 108
Roberts, Richard, 140–41, 157–58
Rockefeller, John D., Jr., 38
Rockefeller Foundation fellowship,
    38–39, 76, 79
Rome, 3, 9, 11, 15–16, 20, 22, 28, 31, 33,
    45, 47, 54, 65, 72, 80, 81, 93–101, 108,
    122, 152, 154, 161, 224, 307
  Allied liberation of, 217–18
  ancient, 7, 8, 106
  Fermi professorship in, 47–48, 56,
    65–69
  March on, 37–38, 48, 72
  physics upheaval of 1936–37, 110–11
  postwar, 286–87
  World War II, 161, 216–18, 287
  See also Via Panisperna

Rome Meteorological Institute, 18
Roosevelt, Franklin D., 156, 235
  death of, 235, 278
  nuclear policy, 158–59, 173, 182, 204,
    235
  war policy, 173–74, 182
Rosenbluth, Marshall, 268
Rosenfeld, Leon, 136–37, 141
Rossi, Bruno, 119, 226
Rotblat, Józef, 235, 237, 244
Royal Italian Academy, 72–75, 83, 88,
    123
Rutherford, Ernest, 43, 82, 85, 94, 97, 111,
    263
  death of, 111
  *Radioactive Substances and Their
    Radiation,* 23

Sachs, Alexander, 156–58
San Francisco Chronicle, 145
Santa Cristina, 57
Santa Fe, 208, 214, 215, 227, 229, 253, 254
Sardinia, 8, 45
Sarfatti, Margherita, 74
Schrödinger, Erwin, 50–52, 264
  "On the Relation of the Heisenberg-
    Born-Jordan Quantum Mechanics
    to Mine," 51
  wave mechanics, 50–52
Science magazine, 299
Scientific Panel, 236–40, 249, 251
Scuola Normale Superiore, Pisa, 16–17,
    18–28
Seaborg, Glenn, 168–69, 173, 179
secrecy, 178, 280
  atomic bomb, 150, 163, 178, 184–85,
    191, 202, 204, 232, 234, 246–48, 249
  hydrogen bomb, 280
  postwar, 280
Segrè, Angelo, 306
Segrè, Emilio, 2, 66–69, 76, 84, 89, 110,
    112, 119, 133, 159, 169, 179, 226, 249,
    285, 298–99, 306
  Holocaust and, 217–18
  last visit with Fermi, 298–99
  in Los Alamos, 215–18, 229–30
semiconductors, 55
Serber, Robert, 211
Sicily, 29, 45, 47
  Allied invasion of, 216

Siegbahn, Manne, 133
Sinclair, Upton, *The Jungle,* 187
Site W, 204, 205–6, 210, 212–14, 221, 222–25, 236
Site X, 204–6, 210, 212, 221, 236
Site Y, 204, 206–8, 209–16, 226–33
slow neutrons, 102–6, 123, 127, 129–34, 150, 152–54, 163, 170, 172, 263, 285–86
  patent, 285–86
Smith, Cyril, 254
Smyth, Henry de Wolf, 248
Smyth report, 248–49, 261
sodium, 112
solar eclipse, 27
Solvay Conference, 52, 88
  of 1927, 52
  of 1930, 52
  of 1933, 88
Sommerfeld, Arnold, 30, 32, 47, 53, 54, 80, 285
  Atombau und Spektralinien, 23, 30
sound, 19–20
South America, 100
Soviet Union, 37, 92, 177
  Cold War, 275, 277, 280–82, 288
  invasion of Manchuria, 240
  nuclear weapons, 278, 282
  spies, 234, 280
  World War II, 177, 235, 237, 238, 240, 241
Spain, 107, 157
  Civil War, 108, 115
spectroscopy, 43–46
Spedding, Frank, 186
spies, 234, 280
square dancing, 228, 270
Stalin, Joseph, 241, 278
Standing By and Making Do, 232
Stanford University, 276
statistical mechanics, 32–33, 46, 51–52, 54, 55, 69
Stimson, Henry, 235–40, 242, 246, 250
Stockholm, 121, 129, 131
Stone & Webster, 182, 183, 188, 203
Strassmann, Fritz, 129–34, 137, 142
Strauss, Lewis, 278
Super, the, 277–82
supernova, 91–92
Sweden, 122, 127, 129, 133–34

Switzerland, 80, 217
synagogues, 34, 57, 121, 217
synchrocyclotron, 265–66, 297
Szilard, Leo, 146–58, 162–65, 171, 172, 179, 180, 220, 240, 251–52, 255–56, 273, 274, 275
  chain reaction and fission research, 146–50, 152–54, 162–65
  Fermi and, 149–54, 162, 255–56
  immigration to United States, 148–50

Target Committee, 242
technology transfer, 275
television, 201, 262
Teller, Edward, 80, 88, 138, 156–58, 207, 224, 225, 226, 255, 257, 274, 276, 289, 295, 299
  hydrogen bomb and, 274, 277, 278, 281, 299
  testimony against Oppenheimer, 289–91, 299
Tennessee, 204, 205
Tennessee Valley Authority (TVA), 205, 277
tennis, 43, 270
theoretical physics, 28, 30–32, 49–52, 78, 81, 114, 263, 269
  beta decay theory, 88–94, 109, 263
  See also specific areas, scientists, and theories
Theoretical Physics Institute, Copenhagen, 119
thermodynamics, 32–33, 44
Thomas, Llewellyn, 69
Thomas-Fermi equation, 69
thorium, 150
Tibbets, Paul, 242
Tinian, 242
Tokyo, 238, 242, 245
Trabacchi, Giulio Cesare, 95–97, 104
transistors, 55
transuranics, 98–99, 105–7, 122–23, 130, 132–33, 166, 168–69
Trinity test, 1–3, 240–42, 247, 271, 307
Truman, Harry, 225, 234–35, 241
  hydrogen bomb and, 278–82
  nuclear policy, 235–43, 246, 248, 251–52, 278
Turin, 29, 75
Tuve, Merle, 157

Uffizi Gallery, Florence, 117
Uhlenbeck, George, 39, 78, 145
Ulam, Françoise, 291
Ulam, Stan, 224, 232, 266, 281, 292, 299
uncertainty principle, 52, 83
unified field theory, 264
United Nations, 285, 302
United States, 48, 54, 76–77, 87, 92, 107, 150
  academia, 76–77, 112, 118–20, 136–37, 148–50, 154, 159–66, 174–204, 254–57, 264, 268–74
  Cold War, 275, 277, 280–82, 288
  decision to use bomb on Japan, 236–40
  Depression, 79
  isolationism, 167
  Jews, 134
  Manhattan Project, 204–43
  McCarthyism, 288–90, 291, 302
  Met Lab, 176–204
  Pearl Harbor attack, 173–74, 202, 237
  physics in, 76–77, 105, 112–13, 135–243, 254–57, 264–74
  postwar nuclear policy and research, 237, 254–57, 268–74
  press, 99, 121, 123, 141, 160
  research and funding for atomic bomb, 151–58, 162–75, 176–83, 184–243
  uses atomic bomb on Japan, 242–43, 244–52
  World War II, 173–74, 177, 234–43, 244–47
U.S. Army, 164, 178, 182, 184, 207
U.S. Army Corps of Engineers, 182
U.S. Congress, 164, 173, 237, 250, 251, 275, 280, 302
U.S. Navy, 151–52, 164
University of Berlin, 128
University of Cagliari, 45–46
University of California, 206, 276, 306
University of Chicago, 133, 145, 154, 171, 174–75, 185–86, 254–57, 264, 268–74, 302, 307
  CP-1 experiment, 184–204
  Institutes of Basic Research, 254–57
  Laboratory School, 214, 301
  Met Lab, 175, 176–204
  postwar physics and, 254–57, 264, 268–74
  Stagg Field, 188–98, 307

University of Chicago Press, 268–69, 296
University of Messina, 29
University of Michigan, 78, 154
University of Naples, 114, 117
University of Pennsylvania, 231
University of Pisa, 17, 22, 23, 24
University of Rochester, 266
University of Rome, 20, 22, 33, 47–48, 120, 122, 287, 307
University of Vienna, 127
uranium, 98, 99, 105, 106, 129–34, 150–58, 163–75
  CP-1 experiment, 184–204
  Met Lab project, 175, 176–204
  neutron bombardment of, 105, 106, 129–34, 137, 140–50, 163
  oxide dust, 187, 190
  research and funding, 151–58, 162–75, 176–83, 184–243
Uranverein, 158, 219–20
Urey, Harold, 160, 207, 254, 255
Uruguay, 100
U-235, 150, 166, 168–73, 181, 183, 205, 221, 222, 236
  bomb, 221, 222, 236, 240, 242–43
U-238, 166, 168, 173, 181, 205, 223
U-239, 168, 223

Vatican, 9, 74, 224
Venice, 7, 35
Via Panisperna, 65–69, 79–80, 84–86, 93–109, 110–11, 117, 119, 152, 153, 162, 179, 215, 229, 269, 285, 307
Victor Emmanuel III, King of Italy, 99
Volta, Alessandro, 53, 66
Volta Congress, 52–55, 285
Volterra, Vito, 34, 45, 48, 75
von Braun, Wernher, 290
von Neumann, John, 231, 273, 274, 281

Wagner, Richard, *Götterdämmerung*, 219, 225
War Production Board, 205
Washington Conference, 138, 140, 149, 157
Washington Evening Star, 141
Wattenberg, Al, 187
wave mechanics, 16, 50–52
Weil, George, 194–95
Weisskopf, Victor, 231

Weizmann, Chaim, 147, 148
Wheeler, John, 137, 223, 224, 278
Wick, Giancarlo, 75, 80, 90
Wigner, Eugene, 147–49, 151, 156–58,
    171, 172, 179, 198, 223–24, 274
Wilson, Robert R. (Bob), 210–11, 225, 247,
    249–51, 256–57, 289, 295, 300
Wilson, Volney, 188, 195
wireless telegraphy, 48
Women's Army Corps (WACs), 229
Woods Marshall, Leona, 60, 180–81, 185,
    190, 193–95, 202, 213, 251, 255, 271,
    279, 281, 299–300
World War I, 17, 19, 29, 34, 37, 47, 53, 57,
    107, 182
  chemical weapons, 107, 238
World War II, 150, 155, 159, 161, 164,
    216–18, 235
  beginning of, 155, 159

decision to use to bomb on Japan,
    236–40
end of, 235, 243, 245–46
events leading to, 115

X-rays, 26

Yalta Conference, 135
Yang, Chen-Ning, 266, 298
Yukawa, Hideki, 264–65

Zeitschrift für Physik, 23, 45, 46, 86
Zinn, Jean, 202
Zinn, Walter, 186–98, 202, 213,
    297
  CP-1 experiment, 186–98
Zionism, 147
Zip control rod, 194–96
Zurich, 31, 80

ABOUT THE AUTHORS

GINO SEGRÈ's previous books include *A Matter of Degrees, Faust in Copenhagen,* and *Ordinary Geniuses.* A graduate of Harvard College, he earned his Ph.D. at MIT and is now an emeritus professor of physics and astronomy at the University of Pennsylvania. He has received numerous honors, including awards from the National Science Foundation, the Alfred P. Sloan Foundation, the John S. Guggenheim Foundation, and the U.S. Department of Energy. Segrè was born in Florence, Italy.

BETTINA HOERLIN taught health care disparities at the University of Pennsylvania for sixteen years. She also has been a visiting lecturer at Haverford College and Oxford University. Her career in health policy and administration included serving as health commissioner of Philadelphia. The author of *Steps of Courage: My Parents' Journey from Nazi Germany to America*, she grew up in the Atomic City of Los Alamos.